The Truth Tells Twice

The Life of a North-east Farm

CHARLIE ALLAN

For Fiona,
with love

This edition first published in 2009 by
Birlinn Limited
West Newington House
10 Newington Road
Edinburgh
EH9 1QS

www.birlinn.co.uk

ISBN 978 1 84158 800 1

British Library Cataloguing-in-Publication Data
A catalogue record for this book is available from the British Library

Typeset by Carolyn Griffiths
Printed and bound by CPI Cox & Wyman, Reading

Contents

List of Illustrations

The Allans of Bodachra: Chae and his elder daughter, Charles and Alexander.

The author's grandmother, Fanny Allan, with Fred Willows (c. 1940), whom she met and married in Canada.

The Heroes of Andover. John Allan and his two army pals who taught him to drink off the piano and throw Yanks to the ducks.

Village hall election meeting 1945.

James Low with the second pair at North Ythsie, 1928.

James and Isabella Low with Joe, Belle, Jimmy, Dod and Albert, 1945.

John Allan admires his elevator and chaff blower, 1948.

Fine new concrete close (needing a sweep already), 1948.

Albert Low with the author in front of the Little Ardo piggery, 1945.

Scotland's first First Minister with the author in the garden of St Nicholas School, 1950.

James and Mary Kelman in Little Ardo's byre, 1948.

Jake Clubb gets the milk away, 1947.

The Hairst Squad 1954. John Allan, the author, Craigie Taylor, Willie Adie, Bill Taylor, Mary Taylor, Jimmy Taylor, Sarge Mackinnon and James Low.

Susan Rennie (née Allan) John Allan's protector, with her youngest son, Benji.

John Allan, united at last with the rest of his mother's family – half sister Fanny, half brother Gordon Willows and two little nieces – in Canada, 1951.

Acknowledgements

The subjects of this book are mostly dead. They are mainly previous generations of my family who told me about their fathers and mothers and about the way things were on Little Ardo. If they had not taken the time I wouldn't have been able to write this, so I am grateful to them for that. An exception is Robin Kelsall, my boyhood pal, or so I thought, who kindly gave permission to use an anecdote from his book Blairlogie Boyhood.

I am grateful to Hugh Andrew of Birlinn for asking me to do a book of anecdotes of the North-east which reminded me that, stuck in the word-processer I had used in the 1980s and quite forgotten, I had perhaps half of what has ended up in this volume. And I must thank his staff who helped me to lick it into shape.

Helen Bleck was the editor who saw to it that what I had written would make sense to people who had not been brought up in rural Aberdeenshire, and Andrew Simmons who oversaw the project, managed the author, not always the most relaxed of men, with a firm hand and calm sough. My thanks to Rhoda Howie, for her imaginative map of the farm, are also warm.

Then, I am very lucky in my family. My three girls, all graduates and aware of the importance of books, all helped. Susie, who is a sub-editor with *Scotland on Sunday*, found heaps of my mistakes and helped greatly with presentation. Sarah, who is one of the current partners in Little Ardo, has a very keen eye for errors, inconsistencies, and non-sequiturs, and found some even after everyone else had been through the manuscript. And Fiona, who was my partner in the family farm and so much else, encouraged, reminded and refreshed me, and cut me the slack which made it all possible.

Charlie Allan
Little Ardo
May 2008

Foreword

'The truth tells twice' was one of the sayings of my maternal grandmother, Mary Yull of Little Ardo, who became Mrs Maitland Mackie of North Ythsie and looked after me during the latter stages of the war. It signifies that, whatever the morality of it, it is wise to tell the truth because if you don't, and you have to tell your lie again, it may come out differently and you may be found out.

But there is for me more to it than that. The saying allows the possibility of occasions when the facts are so horrendously to your disadvantage that you couldn't possibly fail to remember which lie you told. Her saying also allows for the occasions when you remember the facts wrongly. Because if you remember the facts wrongly or mistake what the facts are, then it is almost certain that the same mistake will come out next time and the time after. Such a wrong statement takes on the status of truth – after a while it may even be said to be the truth.

This book tells the truth as I remember it, and it will tell twice if I am the teller.

Charlie Allan,
Whinhill of Ardo,
Spring 2008

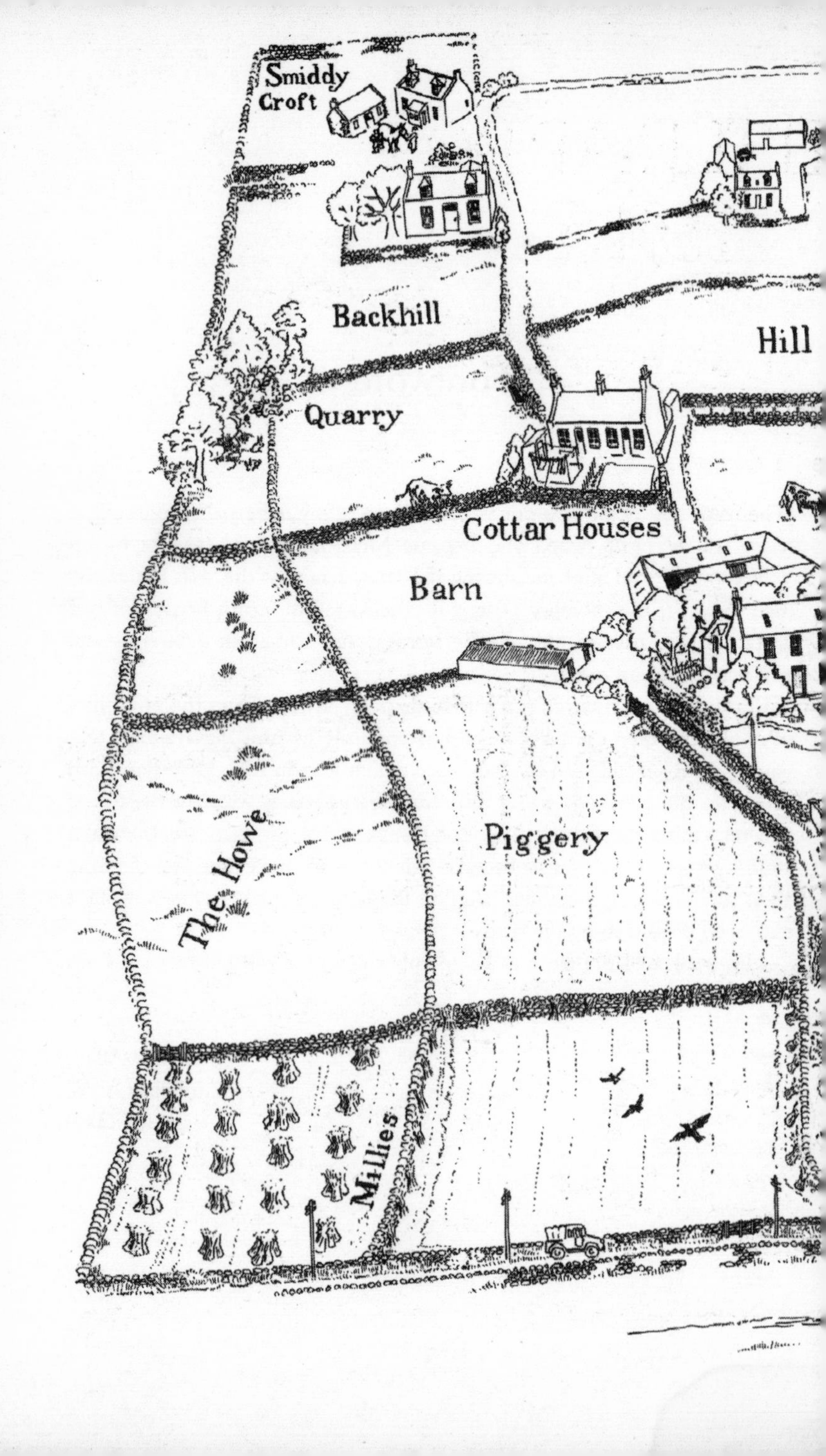
Smiddy Croft
Backhill
Hill
Quarry
Cottar Houses
Barn
Piggery
The Howe
Millies

ITTLEARDO
1945
Loans
Lotties
Single House
Glebe
Auld Croft
Sunny Brae
Three Neukit Park
The Wid
Doctor
Admiral
Black Craig
Waterside

FARMERS OF LITTLE ARDO 1837–2008

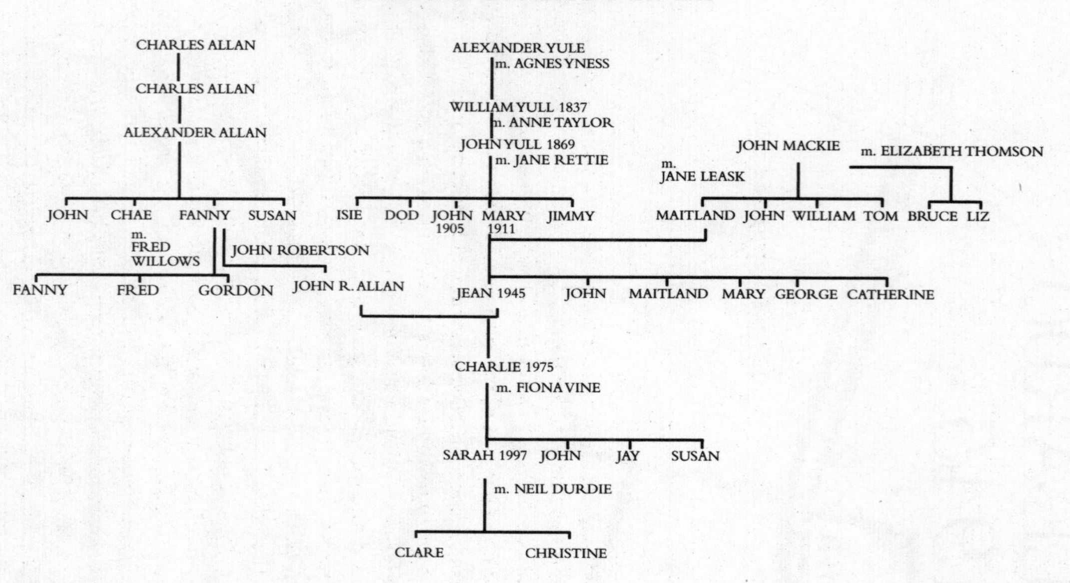

1
The Governess

I was so impatient to get into the world that the taxi didn't even get my mother to the hospital in time. I was delivered at or near the bridge at Stirling where, some years earlier, William Wallace caught the English army divided on either side of the river and put them to the sword in the Battle of Stirling Bridge. Both events seem far off now, but if I was in such a hurry it was surely no wonder. I was to live in a small eighteenth-century two-storey house in Blairlogie, the village that nestles under the southern foothills of the Ochils. The only child (by a series of much regretted accidents rather than by my parents' choice), I was to have two delightful parents all to myself. A large, comfortable mother who wanted nothing better than that I should feel loved, and the gentlest and best-humoured father a child could ever want. He was a freelance journalist, which meant that he didn't have an office to go to and would be available to take me for walks and explain to me the wonders of the world – and, as a creative writer, if there were no wonders to be explained, he could make them up.

Mind you, having a freelance journalist for a father didn't mean you were rich, far from it. When John R. Allan went courting the daughter of a large tenant farm in Aberdeenshire he was given a stern inquisition on his ability to keep the then Miss Mackie in a manner acceptable to her family. It was a situation of which the suitor, a graduate of English Literature, had read many times, but for which he was not well prepared. He was making a very poor job of presenting his balance sheet when my mother attempted a rescue.

'Oh Dad,' she said impatiently, 'he's plenty of money. He gets £5 a week from his writing.'

'Aye,' said my father. 'That was a right good week.'

When you think of that it's a wonder I got here at all. But then, old Maitland Mackie was just playing games. He knew the matter of who his daughter would marry had been out of his hands long before the young couple had gone to see him. Besides, he welcomed the prospect of having a son-in-law who would be able to worry about things other than the price of fat cattle at Maud or whether the Good Lord would give us, for a change, some decent weather for the harvest.

To some extent, old Maitland got the son-in-law he deserved. Though himself a staunch supporter of the Tory party, it was he who, with great care and gravitas, explained the principles of socialism to his eldest daughter, Jean, and to his eldest son, John. It was a very odd thing to do but he did it so well that the old man kindled in his six children a lifelong interest in politics which never led any of them back to the Conservative party, or anywhere near it. I think the old devil did it mostly to annoy his wife, who was a believer in God and, a very close second, the Tory party. The old capitalist told his children how much fairer the world would be if the workers got a bigger share and how much more efficient industry would be if it were run in the interests of all and to a plan, while his wife grew more and more upset. 'Shame on you Maitland Mackie,' she would say, 'Oh, fei, fei!' Mind you, I'm sure she would have managed to forgive her husband had she lived to see her son John take his seat in the House of Lords, even as a Labour peer of the realm, some fifty years later.

Considering that seditious teaching, and the fact that she had studied English Literature at Aberdeen University, it wasn't surprising that my mother should marry a socialist and a writer. But that didn't stop some of her farming family being surprised and even disapproving. When John Allan appeared at a family wedding in a white Moygashel jacket and a red shirt he quickly became known as 'the Ice Creamer'. It is true that the writer who eight years later was to stun an officer selection board by saying that Virginia Woolf was his favourite author, wasn't a natural fit for the Mackie family. If he was more highly developed intellectually than the farmers of Aberdeenshire, physically their noisy conversations frequently went right over the young journalist's head. He not only spoke more gently but was on average six inches shorter than his three farming brothers-in-law. My father once described his role at parties with his in-laws as, 'Down in the forest something stirred.'

If some in her family thought John Allan a less than perfect choice, Jean's two younger sisters thought him a model – in fact they used him as a pattern when they went looking for husbands themselves. From a family of rather raucous agricultural giants, they both married journalists who, like John Allan, were gentle, of no more than average stature, urbane (at least by the standards of Tarves) and proud to call themselves socialists.

There was another person who took to John R. Allan when he came courting at North Ythsie. The local minister was not a member of the Mackie family, but he was a very important and much loved part of it. The parents, who were devout, thought him a fine spiritual leader and the young Mackies liked him because they were not convinced that 'the Revie', as he was nicknamed, believed much of that which he had been called to teach. Certainly the Reverend James Murray had sufficient doubt to endear himself to the six Mackie children who, to varying degrees, found literal Christian doctrine hard to swallow whole.

The Revie and my father became very firm friends. They had much in common including, so the Old Mackies thought, an astonishingly particular interest in the weather. The Old Mackies ran a teetotal household in which there was a bottle of whisky, but it really was for medicinal purposes. Their children, and the suitors they brought home, on the other hand, were all very partial to a drink, in moderation – and not too much moderation at that. When the Revie came to visit, which he always did if John Allan was to be there, the two young men were forever taking a quick step outside 'For a sniff of the night air', 'To see if the wind was still in the east' or to assess the likelihood of snow. And if the two of them had a bottle plonked outside it was no business of anyone's but their own . . . normally.

But on one particular night the two friends had been out many times checking the snow that had fallen all evening. When the Revie decided it was time for him to go home to the Manse to prepare himself for the sermon he would give his flock the next day, John Allan offered to walk him part of the way home. There being still something left in the bottle they took it with them. The chat being good, they had soon covered the mile and a half and my father prepared to return to the farm. But the chat had been so good that the minister decided that he would now accompany his friend part of the way home, so they set out to retrace their steps through the snow.

Half a mile from Tarves on a bend of the road lies the bridge over the Sonah Burn. It was a favourite meeting place for the young hopefuls in those days, and a natural place for a pause before tackling the brae up to the village. There the minister and the journalist relieved themselves of the product of a long night's surveillance of the weather. And being in high spirits they wrote their names in the snow.

By now it had stopped snowing and a hard frost had set in.

The faithful of the parish on their way to Sunday service could clearly see that two people with big feet had been writing in the snow. Sadly the Revie had the steadier hand. While my father's name could only just be made out, the faithful could clearly see their minister's name written in yellow. Of course there was more than one James Murray in the parish of Tarves in those days and there were those who said it was scandalous of whoever it was who had written the minister's name like that.

So, in 1934 my father and mother did get married and when in the fullness of five years I arrived they set about giving me a childhood that would be full of protection, warmth and humour. They were doing a grand job when Hitler went and spoiled it all. In 1939 he provoked war with the civilised world and, as the Americans were late again, the might of my family was required to stop him. The young writer became Sapper Allan of the Royal Engineers and I was left, at first, with my mother. In 1941, when the lights had gone out all over Europe, she and Margaret Wilkinson, a childhood friend of hers, took the old House of Mark up Glenesk in Angus, and while she renewed that old acquaintance I acquired my first girlfriend in her daughter, Anna. Anna's father had also been called up and the two young wives had fled up the glen to escape the rape and pillage that would surely follow the German invasion which they were both convinced was inevitable.

My mother once told me (and subsequently denied having told me) that after a while of their own company and the quiet of the hills, the horrors of conquest started to seem exaggerated. And that by the end of the first month the two young wives were taking turns going down to the head of the glen to see if there were signs of any Germans yet. Meanwhile, Anna and I played interminably by the banks of the Esk in what seems now like one long summer day, oblivious to the horrors that were going on in Europe and the imagined horrors to come. In 1941 our parents' mood often came near to despair. Anna's mother, on one occasion, cried out, 'Surely Churchill has a plan', to which my father, home on a brief leave, replied sourly, 'Aye. Hot-foot for Canada.'

It stands to the eternal credit of Anna's dad and mine that Churchill's contingency plan was never required. The news from the fronts improved and the threat of rape and pillage on British soil receded. The two young wives decided it was safe to take their children back to their homes. We returned to Blairlogie.

There for a short time we enjoyed village life even without John Allan. Blairlogie has been spared the sort of suburban development that has spoiled most of the villages that are near to the centre of things. For whatever reason, the developers have been unable to get to work and overwhelm the place, despite the fact that it is ideal for commuters to Stirling, and even Edinburgh would only be half an hour away were there not so many other people trying to get there. So you can see Blairlogie in 2008 more or less as it was in 1942.

Though the amenities only amounted to one church and one wee shop selling groceries and anything else you might order and be willing to wait for, we were only a couple of miles from Bridge of Allan or Stirling. It was a trip we seemed to make often. It was a bumpy ride on a single-decker bus with a clippie, Big Bella, who always smiled at the little boy who was pretty enough to have been a girl. We would go a few hundred yards along the Hillfoots road and then climb steeply and windingly up past the Wallace Monument while my mother told me all about the betrayal and suffering of William Wallace and I wondered at the size of his sword. Then down the other side into another world of traffic and bustle, though as the war proceeded there was less and less in the shops. It must have been a good deal later that she told me about it, but my mother claimed to have overheard two ladies gossiping on that bus, about a woman who was well past her time. 'They'll need to put in a ferret, afore that wean'll come oot,' said one gravely. 'Aye,' said the other, 'or put her on the wee bus to Blairlogie.'

I was really too young to remember much about the village except its quiet and its friendliness. I played outside most of the time. My mother reckoned I was quite safe from all the things modern mothers fear, even from traffic, so long as we didn't stray down to the main road, which took several cars an hour from the Hillfoots towns of Alloa, Alva and Menstrie to Stirling. Many of the fathers were away in the forces but there were plenty of children to play with, of whom only Robin Kelsall and Jim Clark had names that I can remember. I can still remember the breathless excitement with which I came in to tell my mother, 'Jim Clark's cat's had kittlins. There's one black and one white and one perple.'

We saw almost nothing of the war. There was one day when a convoy of army vehicles stopped for what seemed like hours on the Hillfoots road. It stretched as far as we could see in both directions so it must have been at least half a mile long. There were tanks, lots of lorries and troop carriers. The soldiers were very friendly and, no doubt missing their own homes and families, took us bairns in hand and gave us a day to remember. We swarmed about over all this deadly hardware and got sweeties.

The other clear memory I have is of looking out of the window across the square to Jim Clark's house and seeing a very urgent coming and going. I didn't know what was happening but I could see it was important. I learned later that some lads from Alva had been playing up on the Dumyat Hill behind the village and that they had come across some unexploded ordnance left behind by the Home Guard, who used the hill for training. It exploded and one lad had been badly hurt. What I had seen was him arriving back at the village. The lads had been afraid to raise the alarm because they had been promised such a hiding if they went on the hill in their Sunday boots. The implication of my mother's story was that the boy had died when he might have been saved and that it showed how important it was not to terrify children into doing what they are told. We must have been able to hear the bombing of the Forth Ports but I was blissfully unaware, happy in my village full of friends and in the easy-going but loving care of my mother.

But the idyll was to be shattered. That cruel war, having demanded my father, now required my mother. She left to aid the war effort as a teacher, and I acquired a governess. I have no doubt the governess was in her own way a good woman but all I can remember is a middle-aged monster.

Though I had been parted from Anna and my parents I still had my idyllic village and lots of children to play with, though they were, like almost the entire world, a bit older than me. We played skipping, singing and ball games, and cycled for hours down the village street along the Hillfoots road towards Menstrie and back up the little bumpy lane which led from the post office back up to the top of the street. It was in that cycling that I suffered one of the few frustrations of my young life. While the rest of the kids swished round the circuit on their bikes, I was always half a lap behind on my three-wheeler. However, I was a resourceful lad even then and thought of a way to pep up my trike. From the glory hole at the back of the stick shed I produced a pot of the pink paint, with which my mother had just freshened up the old house. I had no brush, mind, but I managed fine with my hands and soon had most of the trike and a considerable area of the front step, and myself, gleaming pink and new. You cannot possibly imagine the pride of achievement with which I sought my governess to show her my handiwork before launching my now supercharged trike on the rest of the village.

To cut a long and painful story short (for my sake not yours, for it pains me still to tell it) Miss Dread (I'll have to call her something for I don't even know if she had a name) was neither favourably impressed, nor was she amused. To my complete bewilderment she gave me the very mother and father of a hiding. She tore my clothes off, hauled me inside, scrubbed me hard with something nippy and bunged me into bed. Then she slammed out of the room only to reappear as though in an instant carrying her fur cape. It was one of those whole foxes that ladies used to wear before the days of wildlife conservation and political correctness. It had whole feet with the claws still on them, a stuffed head with glass eyes and sharp teeth bared and ready for action. She thrust that evil face into mine and screamed through clenched teeth, 'And if you ever do anything like that ever again this fox'll come and eat your head right off.'

Already bereft of my infinitely patient parents, the quality of my young life suffered a serious decline. Every time I went to do anything – good or bad, for I had only the haziest notion of what might be the difference – I had to weigh the prospect of being devoured by that fox. Often I slept under the covers in case, in the night, he should come to bite me.

I am sorry that I can't do justice to Miss Dread. I'd like to for there is no bitterness left, though to this day, sixty-four years later, I still don't like dogs. But as real as she is to me today, when I try to envision her, all I see is a middle-aged female figure with an appalling enraged fox's face blazing hatred at me from under dishevelled grey hair.

I am glad to say that my misery was shortlived. On one of her leaves I got the chance to tell my mother of my terror, and took it. She didn't let me down. Miss Dread was sacked and she left without a reference, while I was packed off to the farm in Aberdeenshire where my father had made such a poor job of asking for Jean Mackie's hand in marriage. There I was to stay with my granny, and that was another and infinitely happy story.

2
The Refugee

I moved from the village of Blairlogie on the fringe of Scotland's industrial central belt to the care of my grandmother at North Ythsie farm (pronounced 'I see') about twenty miles north-west of Aberdeen. And so it was that I entered the farming community of the North-east of Scotland.

I could hardly have chosen a more agreeable point of entry. I could have been one of the cottar family of fifteen who shared a two-bedroomed house at North Ythsie, or the family of seven who, as late as 1949, still lived under a tin roof and on an earth floor on one of the county's famous farms. But no. I joined the household of Maitland Mackie, the tenant of 400 second-class acres who, when so many farmers were failing in the 1930s, had the skill as a farmer and the nerve as a borrower to buy up farms all over the North-east, including two in the fat lands farther south: Bent, perhaps the best farm in the Howe of the Mearns, and at the end of the war took Benshie, one of the best in the Howe of Strathmore. So the Refugee was to have a comfortable life.

The Mackies had had six children but they were all flown, leaving empty rooms all over the old farmhouse. The Old Mackies were glad to have an infant back in the house, or at least they certainly made me feel that they were. Mind you I did get a terrible shock when, on one of my explorations of my new world, I found that my granny had one of those confounded fox capes. I think she must have known about Miss Dread for she soon had me convinced that, as real as its teeth surely were, the fox was indeed quite harmless and wasn't even very warm. I remember being puzzled as to why, in that case, she didn't throw the damned thing out.

The farmhouse at North Ythsie was in two halves. There was the rather grand Victorian half, built from the profits of the Golden Age of Agriculture in the third quarter of the nineteenth century, and there was the older and more modest eighteenth-century farmhouse set at right angles at the back. The ground floor of that housed the gardener, whose wife fed the single men. My grandparents, Lizzie the maid and I had the rest to ourselves.

It was a split-level affair because the original was half a storey lower than the new, and that gave my grandparents' house a rambling effect. If you went in at the back door you could go downstairs to the cellar or upstairs to the ground floor of the new house, from which you could go upstairs to the first floor of the old house. And if you went up again you would come to the first floor of the new house. Even then you could go up again to the attic of the old house. So it was a six-storey skyscraper to the small boy in 1943.

My room was on the fourth of those floors, where every evening I went to sleep to the sounds from the close (as we call the farmyards in Aberdeenshire) of the older children playing the games of impending adolescence. They seemed to come from a world of great romance but, knowing that I would one day be old enough and that it would be a waste of time asking, I was content to look out or listen and wonder. My favourite was Rita Davidson, the spirited and precocious daughter of the grieve (working manager) of South Ythsie, the farm which glowered at us across the Ellon to Tarves road. She had long dark hair and a body which was filling out though she could have been no more than ten. Others noticed too and I will never forget my anguish as I watched two of the bigger boys, an arm apiece, spinning her round and round. It looked as though her arms must surely be torn off, and in my innocence I mistook Rita's squeals of ecstasy for screams of agony.

Every morning I awoke to the crashing of the ten-gallon milk cans being loaded onto the lorry for Aberdeen, followed by the slow crunch, crunch of the heavy tackety boots as Walker the milky and the single men came across the close to the kitchen below for their brose at seven.

But the young refugee billeted with Maitland and Mary Mackie didn't have brose for his breakfast. He didn't even know at that time that brose was just two to four tablespoonfuls of oatmeal and salt stirred together with boiling water from the kettle. Brose is a sort of porridge for men in a terrible hurry. The men used to pick up their brose cap, spoon in the meal, walk over to the fire for the kettle and pour in the water. They then stirred it up with the handle of the spoon and sat down to dine all in one movement. Even if they did get tea as well, the men could be out again in five minutes.

War or no war, up half a floor and an hour later, my grandparents and I had bacon and egg. At least I can't remember for sure getting bacon, though I have no reason to doubt it, but what I do remember with mouth-watering clarity was my greatest joy in gluttony. That was to go round the table after the meal and carefully clear all the plates of the delicious white fat that everyone else had left. I didn't like the rind so well but, with characteristic thoroughness, I ate that as well. As my granny explained to a fellow guest who had indicated surprise at this procedure (or perhaps at my getting away with it), 'There's no use keeping a pig and Charles too.' Indeed, Mrs Mackie didn't keep a pig, though her husband kept them by the hundred.

There were lots of visitors in 1944 to wonder at my gluttony, for my grandparents were excellent company and most hospitable. In particular they welcomed a string of young airmen from Australia who, like me, used North Ythsie as a home from home. I still meet one, John Yull, my granny's nephew who came once or twice, but my favourite at the time was called Russ. He took me sledging and walking up the hill to the Prop of Ythsie, a granite tower erected in 1862 to the memory of the Lord Aberdeen who had been prime minister of Great Britain at the time of the Crimean War. From there we could see the sea. I adored Russ and admired him particularly for the fact that he didn't have a hairy face like nearly all the other grown men at that time and in that place.

They never gave me a proper answer to my questions about when Russ was coming back. He had promised me that he would come back and I was puzzled and hurt that he should only come the once. Surely he had enjoyed being at North Ythsie and would have liked to come back? I was in my twenties when I realised that of course Russ would have come back had he made it back from Germany . . . and I was in my forties before my mother told me at last that he had been shot down in his Lancaster bomber. My granny it was who used to tell me the importance of telling the truth, with her emphatic saying, 'The Truth Tells Twice'. She lived by rules, and telling the truth was one of them. I suppose what she kept telling me about Russ coming back 'next time' was the exception that proved her rule.

My grandparents' discipline was quite different from Miss Dread's. Indeed, the matter of discipline just didn't arise. There was no question of doing anything of which they would disapprove. If either of them ever reprimanded me it was done in such a way that I didn't feel bad.

The day had a rigid routine. Eight o'clock was breakfast. Ten o'clock was coffee and cheese biscuits. One o'clock was dinner. Three o'clock was afternoon tea with pancakes and strawberry or raspberry jam and maybe a scone. Six o'clock was supper and seven o'clock was bedtime. That was every day without failure or argument. The week had a routine which included Fridays to Aberdeen for the mart, business or shopping. And I didn't like Sundays. It wasn't the going to church. That was all right, with lots of ladies to give me a sweetie and say how pretty I was and that I should have been a girl (it was only much later that I realised macho men didn't think that was much of a compliment). No, what I didn't like about Sundays at North Ythsie was that there wasn't supposed to be any work done. My granny interpreted that as meaning only sandwiches, made the day before, for Sunday supper. Many a weary traveller, used to the excellent meals at North Ythsie, suddenly felt much wearier when he realised that he had done it again and arrived on a Sunday.

Tiresome though I found this Sunday regime I accepted it with a good grace. And that was more than the six Mackie children – my mother, three uncles and two aunts – had done. They found it increasingly difficult as time wore on to reconcile their desires to enjoy themselves with their parents' desire to observe the Sabbath and keep it Holy. The row that got up when the second son chose a moment of silence at dinner to ask his little brother, in a loud voice, how he had enjoyed his outing to the annual horse fair and Sunday carnival at Aikey Brae was savoured by all those who were there until the day they died. But more typically and more revealing of how the family operated was a conflict which ran for years. I don't suppose Mrs Mackie would have let her husband build the tennis court if she'd realised what trouble it would lead to. What the Mackie children had done with most of their Sabbaths was to break them discreetly where their parents wouldn't see. But the tennis court was just outside the window the old man had put in to let more light into the sitting room.

'What,' the children asked, 'is the difference between going for a walk on a Sunday, which is allowed, and walking round the tennis court after a ball, which is not allowed?'

Well, the old folks managed to handle that one all right. When you go for a walk it is peaceful and you can contemplate the wonders of God and His works. Whereas when you are playing tennis you are racing around trying to think of ways to win and arguing at the top of your voice about whether the ball was in or not. And who had won and who had lost. But Sunday was such a good day for tennis because that was the day when you weren't allowed to do anything else and neither were your pals, so everybody was available for tennis on a Sunday. Eventually the parents were worn down into an uneasy compromise; tennis would be allowed on a Sunday but they would not be allowed to count the score.

When my uncles and aunts left home the old folks breathed a sigh of relief and let the tennis court fall gently into disrepair. By my time during the war only the keenest tennis player would have dreamt of playing on it on any day of the week. Next to the tennis court was the best garden among the many I have enjoyed. It was two thirds of an acre in extent, and while there was a small rose garden at the house side, and a bed or two of pinkies and anemones that my granny picked for the shop, almost all of it was given over to food production. There were all the usual vegetables but, and it is a very big but, apart from the peas, two other things interested me about the North Ythsie garden: the first was the fruit. There was an abundance of fruit. There were always two strawberry beds. The first and smaller bed held the young bushes, which had their flowers removed so they could concentrate all their efforts on producing foliage to make a super crop the next year. The bigger bed, of mature plants, produced fruit for immediate consumption. It was under a net supported by a rickety frame whose uprights were paling posts, some with extensions on them. That protected the crop from all birds but for a few determined blackbirds who often paid for their feed with their lives as they got hopelessly entangled in the net trying to get out again. Like the blackbirds, the young refugee was not allowed under the net but could always get the odd fruit that had been missed among the young plants or which was reachable without going right under the net.

There were several lines of beautifully staked raspberries also under netting, but again, I could always get a few berries from the young plants which had no right to bear fruit. There would have been perhaps thirty or forty bushes of black- and redcurrants, and gooseberries. Of those, the gooseberries were a particular favourite and, although there were nets for those also, it seemed to be possible for me to get a bellyful of those. Mrs Mackie was very proud of her fruit and she pulled out all the stops when, much later, in 1947, Tarves started its flower show. I hesitate to say that she cheated, but she was early in to the sort of tricks gardeners play in order to win the tiny prizes and huge honour at their local flower and root shows. One of the many gooseberry bushes was stripped of all but a few handfuls of its fruit when they were tiny, but after she could see which were perfectly formed and disease-free. That left the bush to push all the goodness of half a barrowful of muck into the production of a few of the most enormous, regular-shaped and blemish-free berries. The best of those were to be put out on plates and entered for the show, where my grandmother was quite sure they would win first prize.

Once, a couple of days before the planned harvest of those berries, I visited my grandparents. After a short session of asking me about everything I was dismissed and made straight for the garden. The peas and the gooseberries were the targets. I found the peas first and had a lovely feed of those. Oh how I was to wish I had left it at that. But no. I must have a gooseberry. I remember well seeing this bush which had hardly any berries. On the other hand, those that there were, were the biggest and juiciest I had ever seen. It was what I suppose people nowadays would call 'a no-brainer'. I ate the lot and proceeded to fill up from the other bushes which were hanging with an abundance. I remember that the smaller fruit were sweeter. I even remember thinking I had made the wrong choice. But I had no idea how wrong the choice had been. It would be easy to say that I was soundly beaten – even the most politically correct do-gooder would surely understand why – but I was not abused in any way. All the same, it was the only time I have seen my granny distressed. All she did was explain that she had been preparing those bushes for the show and how she would not now be able to get a sufficient sample from the remaining, undoctored crop. I still feel bad about that and whenever I am stealing fruit, which is quite often, for there are no fruits that taste as well as those straight off the bush, I always avoid the biggest berries, just in case.

I would guess there was ten times what the family could possibly have eaten in Mrs Mackie's garden at North Ythsie, even though much was made into the jam that so enlivened the afternoon teas. The surplus was grown for the Mackie's Aberdeen Dairies shop in the town and for Mrs Mackie to take as 'a minding' wherever they went visiting. I liked all that fruit when it was put on the table at dinnertimes and suppertimes, when it was served with whipped cream. My grandfather took his strawberries with pepper, swearing that it brought out the flavour. I thought that a very grand idea though I never tried it then. For all that, in my age, I have come to take a little pepper, though I think it is not so much to bring out the flavour of the berries as to bring back the flavour of North Ythsie in the great days.

There were also apple trees which produced excellent cookers but nothing to interest a small boy, and pear trees, which seemed not to get enough sun, for the fruit was small and hard and tasteless. But the greatest joy in the garden were the two large and productive Victoria plum trees. They were magnificent, and so sweet. For a few days in October I could have lived on nothing else.

The second thing that made the North Ythsie garden important to me was the hedge that surrounded it. No doubt attracted by all that fruit, and all the dung Mr Walker put on, and the flies and worms that it promoted, the hedge was always full of birds' nests. Not only that, but as it was trimmed to about six feet high along two sides of its length I could access all the nests I could find, and I found them all. From April till the last nestful was fledged in July 'round the hedge' was part of the daily routine which made life at North Ythsie feel so secure. After I had finished the bacon rinds, which didn't take long, I was off round the hedge to see what I could find.

The range of nests wasn't great. There were blackbirds, chaffinches, green linties, hedge sparrows (it was not until very much later that I found the much better name, dunnocks), ordinary sparrows and the occasional house sparrow, though they mostly availed themselves of the ample nesting opportunities in the steadings. I once found a wood pigeon and at least once a blue tit. I became such an expert that I could anticipate where the next nest would be. The birds spaced their nests out as far as was practical. That meant that at the height of the season the nests were perhaps seven yards apart. And as soon as one lot fledged you would be sure that nearby another bird would start to build. And if you 'ruggit' (pulled down) the old nest another would be built very near and even on the same spot. The talented birds' nester knew the thick bits of foliage in which the birds thought their nests might go unseen, and there were only so many of those.

But the hedge was not the garden's boundary between the road on one side and the fields on the other two. There was, as there still is, a drystane dyke outside it. That left about eight yards for trees and anything of which you wanted to dispose; garden rubbish which wouldn't make good compost because it was too fibrous, like prunings, or too acid, like potato shaws. It was in a pile of the brushwood that was used for staking the peas that I found, in this no-man's-land, my first yalla yite's nest. There were five eggs with markings that looked like a map which, if you could just unravel its meaning, would lead a boy to buried treasure. The trees in no-man's-land had grown too tall and had been keeping the light from the garden, so they had been cut off at about seven feet. The cuts hadn't been treated, so this produced hollow trunks that sometimes filled with water and sometimes filled with nests. I found an owl's nest there, and several jackdaws'. It was in one of the many tin cans that were tidied into the woodie surrounding the garden that I found my first robin's nest.

The attraction was the eggs. In those days nearly all small boys who lived in the country collected eggs. We pricked the ends and then, holding the thick end to our lips, we blew out the contents so that the eggs could be stored without going rotten. Mine were kept in a cardboard box with cotton wool in it and set out in display with the pheasant egg first, then the wood pigeon, then the partridge. I ached for the day when down at the bottom of this progression I could display the egg of a Jenny wren.

At first it seemed an easy job. With most hedge birds laying five eggs in a clutch the box would soon be full. But Mrs Mackie imposed rules. I was only allowed to take one egg from each nest. I now think that was a good rule as almost no bird could keep up with feeding five young, so removing one egg enhanced greatly the chances of the other four. And what would the attraction have been in boxes full of common blackbirds' eggs? But the trouble, from the North Ythsie birds' point of view, was that I was by no means the only collector touring the hedge. I even tried my granny with, 'but if I dinna tak aa the eggs the Crombies'll jist tak them.' I got nowhere with that one and the collection grew only slowly. Later my freedom as a collector was further restricted. I was only allowed to take an egg for my collection if I didn't already have one of that species. As there only were perhaps a dozen different birds nesting in my hedge and even in my woodie that was restrictive indeed, but it didn't dim my enthusiasm for bird's nesting. My grandfather told the laird about my passion, and he got me an oystercatcher's egg and even gave me a permit to range over the whole Haddo House estate in search of eggs. I have it still. It says: 'Master Charles Allan has permission to pass through the Haddo House Policies during the pleasure of the proprietor for "bird's nesting". This card must be shown at the Lodge Gates and on request by authorised persons.' It is signed plain 'David Gordon', who had returned from the war a major and who had not yet succeeded his uncle as Lord Aberdeen.

Maitland Mackie did most of his work by telephone. It was by his chair on the right-hand side of the fire in the large sitting room. On the mantle was the clock that his brother-in-law George Yull had made, ticking noisily between the two black folk carved in ebony that their son George had sent from Africa. There he sat reading and reaching round to the desk for a used envelope on which to do his cash-flow projections, or for his notebook. This was an important management tool, for Maitland had an ingenious method of keeping his telephone bills down. He had a page in the notebook for each of his sons and for all the other people who were likely to phone him and, instead of phoning them up every time he had an order to give or a request to make, or when he had anything to tell, he waited till they called him. Then he was ready with his list. It was a system that required a high level of *sang froid* and organisation, but then, he had plenty of both.

Maitland Mackie depended a great deal on the phone and it was a machine he understood well. On one famous occasion he was discussing with the laird the delicate matter of taking over the title deeds of one of his rented farms when he broke off negotiations with, 'Aye, mi-lord, I don't think we should say any more. You see the lady at the post office has a habit of listening in.'

'I do not listen in,' came an angry voice, 'and shame on you Maitland Mackie for saying such a thing and you an elder of the Kirk.' (Which he was, for sixty-nine years.)

Despite his business interests, which extended well beyond farming to company directorships, college governorships and the Presidency of the

Farmers' Union of Scotland, my grandfather always seemed to have, in those wartime years, plenty of time for me. He even contributed to my political education in much the same way as he had done with his two eldest children twenty-five years earlier. Except that while he had taught socialism to my mother and my uncle, my conversion was the other way round.

As a good socialist who believed that most of the troubles of the working classes in the 1930s could be brought to the door of capitalism, my father had taught me, when home on leave, the old socialist rallying cry, 'Workers of the world unite. You've nothing to lose but your chains.' That stood me in good stead with his bohemian friends, but, as you can imagine, it went down like a lead balloon among the prosperous farmers of Aberdeenshire. I will never know if his motivation was to help me socially, to guide me politically or just to annoy his son-in-law, but Maitland Mackie soon put that right. Whenever there were folk in, which was very often, I would wait for a lull in the conversation and say in a loud voice and with the conviction of the converted, 'I believe in private enterprise', and thump my fist down on the arm of a chair. What with being so pretty and only four years old, that went down really well; it was a huge improvement on 'Workers of the world unite'.

As well as my infinitely patient grandfather and my strong and constant grandmother, and the stream of young airmen for whom I was a surrogate little brother, I had more love heaped on me by my grandfather's two ancient maiden aunts, the Misses Gibson. Bessie and Barbara completed my conversion by giving me the wherewithal to become a practising capitalist. They gave me £1 at Christmas, bade me put it in the Savings Bank and explained that by next Christmas it would be worth with interest £1 6d. And indeed, so it was.

It wasn't just because they gave a start to my savings (which were one day to make it possible for me to go on a cycling holiday in France with a rather nice Dutch girl) that the old ladies were important to me. They fascinated me with their slightly faded silks and velvets, and their brooches and necklaces which took up so much of the dressing-table in their room on the fifth floor of the North Ythsie farmhouse. They were the only women in my entire world who were neither young nor had a husband ... not even a dead one, like old Mrs Fyfe at the Sonah Bridge. They were also the oldest people I knew, and I thought it was perhaps the lack of husbands in their lives that made them so enjoy running and hiding round the six floors of the farmhouse and playing cowboys round the mountainous furniture in the sitting room. It certainly never occurred to me that they might be working hard at keeping me amused.

The aunties were about eighty in the early '40s but they were as spry as you like and I felt it was a duty upon me to keep them amused. On one occasion when my grandmother insisted on taking me with her to Aberdeen, I apologised to Bessie, the younger of the two, for having to spoil her fun but bade my grandfather play with her until I came back 'to keep her oot o' languor'. And I once showed Barbara a card trick that left her wide-eyed. She looked at me across the best part of four-score years and told me that truly it must be magic. I hadn't the heart to tell her that it was mere dexterity that had brought all the Knaves to the top of the pack, but when I went off at last to my bed I heard Barbara tell my grandfather, 'You know Maitland, I remember teaching that trick to your grandfather sixty years ago.'

Also part of the household at North Ythsie when I was a refugee was Lizzie Cruickshank the maid, who had what had been the front room in the original farmhouse. She was what people called 'a poor thing'. She worried about things unnecessarily and indeed she worried so much that eventually poor Lizzie started to lose her memory. When she was with her next employer she went missing, was found sleeping rough and finished her days in a mental hospital. But when I met her in 1943 Lizzie was in full command of herself and a tireless and willing servant who did all the rough work about the house but could also cook and serve at the table when required. Looking back, Lizzie was a physically unattractive, large woman with moles and hair growing on them, but to me her fat was just more warmth and affection. Her system of child management consisted of letting me have everything I wanted. Despite the wartime shortages she once let me have about two ounces of margarine on my boiled potatoes. And she took me on the back of her bicycle the six miles to Ellon and gave me such a feed of icicles (ice lollipops) from her own meagre wages as to make me sick and get her a row from her employers when she brought me home. It was Lizzie who introduced me to serious poetry and just how serious it would have been for the maid had my granny found out I understood well – even then:

Here's to the oak that stands in the wood.
A standing tot does a woman good.

Even at the tender age of four years, and without anyone telling me, I knew that that was not the sort of poetry of which my grandmother would approve. And I liked Lizzie even better for teaching me it . . . it gave us a secret.

Completing my grandparents' household though not quite of it were the Walkers; the gardener, Mrs Walker and the three of their children who were still at home. They occupied the kitchen and what had been the servants' quarters of the original house. Going down to the Walkers' house was for me out of bounds, so I could only go there by invitation.

That was just as well for had it not been so I would have spent a great deal of my time there. In contrast to all the space and tranquillity in the rest of the house, the Walkers' domain was always humming. In addition to the three daughters and the occasional lad, there were the single men who slept in the bothy and came three times a day to be fed, and the place smelt different. There was an earthy smell of hard farm work, Capstan Full-Strength cigarettes and of course, Mrs Walker's frying pan. That was a thing of great wonder as it sat on the old range, which Mrs Walker rubbed up with black-lead every morning. And it was at its most wonderful when it was frying sausages. They shrank as they cooked and fat oozed out of them until they were almost being deep-fried. I spent hours drooling over those heavenly morsels and never had to risk disillusion by tasting one, though I did get the occasional crumb of the crusty bits that grew out of the ends of the sausages as they cooked. What I did get quite often was some of the fried bread with which Mrs Walker soaked up all the fat after the sausages were on the plates. Indeed, looking back at that and the bacon fat at breakfast time, it's a wonder I didn't have heart disease by the time I was five.

Walker (not Mr Walker or Jimmy Walker, mind you), as well as being a good if deliberate gardener and occasional chauffeur, was good with his hands and something of a folk artist. He used to make those jolly fat Clydesdale horses cut out from plywood and painted blue with black and white harness. They were hung on the walls and the Walkers had one as the front of a letter rack. He also made jointed stick men that jumped about and did acrobatics if you manipulated the stick right. Then there were elaborate decorations made out of hundreds of fag packets . . . Craven A, Players Navy Cut and Capstan Full Strength, in great rosettes which could be more than a foot across.

The youngest Walker was a beautiful girl with long golden curls. Violet was twice my age when I arrived at North Ythsie though I did catch up a bit, but she always responded positively to my voice at the door asking, 'Is Bilet' (to rhyme with 'bile' yer heid, my best attempt at 'Violet') 'comin oot tae play?' Like me, Violet didn't seem to get out to play in the close at night and I am glad about that still. I don't think I could have stood to watch the big boys being rough with her.

It is hard for me to believe, with all the security by which I was surrounded, that there was a war on and that my parents were away doing their bit, but at least my grandfather was safe. His bit was producing food for the war effort. This time he didn't even join the Home Guard. He did that during the First World War but he hadn't been a success. He had done quite well at the square-bashing and he was a good shot with a rifle, but when they went on manoeuvres to Haddo House, three miles away, he couldn't stick it. Living in a tent just wasn't the standard of comfort to which Maitland Mackie was accustomed and Mrs Mackie was astonished to find him back in the twin bed beside hers when she awoke in the morning. He never went back.

That meant Maitland Mackie was available throughout the war to show me the wonders of his farming empire. Those included what I remember as a daily trip to the piggery to see how things were going there. To get to the piggery we had to go round the back of the house and through the close to what had been virgin land before the wooden piggery was built between the wars. What a dungeon it was! An ill-lit central corridor had doors off it leading to pens holding perhaps fifty pigs each, at various stages of maturity. When we entered each pen in turn and put on the light the pigs bolted for the small opening which led to their outrun. The inside was dusty and the outrun was dirty. I see English pig farms which remind me of my grandpa's pride and joy, but if there are any such piggeries left in Scotland today, there are very few dungeons like that. Perhaps the old man went so often as a sort of penance. Every week each penful was weighed to give figures from which he could calculate the rate at which the pigs were turning feed into pork. Those figures were posted on the wall as you entered the piggery and showed how near the old man was getting to his ambition to achieve a conversion rate of three – three pounds of meal making a pound of pig. It would never do today when everyone is getting close to a conversion rate of one, but his was a very scientific approach back then.

We also went on what I remember as a weekly basis to Woodlands, his farm five miles away at Udny, in his Riley 9 motor car. The main interest there, at least as far as I was concerned, was the poultry sheds. I particularly liked the incubator house where, if you lifted the right lid (the lids were a bit like the insulators on top of an Aga cooker), you could see the little feathered balls cheeping their first.

I wasn't the only displaced person at North Ythsie in 1944. We had a succession of Italian prisoners of war who lived in a camp at Udny Green and came to the farm to work. They were very popular around the farm. If you gave them a silver thrupenny piece they would make it into a silver ring for you and their singing around the farm was wonderful. My bedroom at North Ythsie was the first place that I heard the 'Bonnie Lass o' Fyvie' but it was also the first place that I heard 'La Donna è Mobile'. They were most generous, with so little to give. One even gave me my first fag, though I have often wondered if that might not have been some sort of revenge for the humiliation his country had suffered in the war. I took that into Lizzie's room and managed to light it off the electric fire. I puffed it a few times. The room began to spin. I was violently sick. Perhaps the kindly Italian was trying to put me off smoking for life and, looking back on how that first fag made me feel, I find it extraordinary that by the time I was twenty I was smoking them up with the best.

Incidentally, that electric fire was the scene of my first experiment. I had been warned many times that if I let water anywhere near the element of that electric fire I would get the most terrible shock. Never one to accept advice uncritically I put this to the test. Forget the risk Benjamin Jesty took in the 1790s when he tested his theory about vaccination against smallpox on his wife and two sons. That had been nothing. I urinated on the red-hot element. To the great satisfaction of the five-year-old scientist, my granny's advice was proved to be quite wrong. I have no doubt that she believed it but what she had told me was not true. I don't know where I got the common sense not to tell her of her mistake.

3
Empty Rooms

My room at NorthYthsie in 1943, although I shared it with no one, was called Mary's Room. It was the second one along the corridor of what was in effect the third floor of the original farmhouse. Farther along was the bathroom and, at the end, the room where my grandparents slept. I had a three-quarter bed all to myself and a reading light with a pink shade and a fringe of goldie tassels that was hooked to the headrest. It became quite unstable when I played about with it – I had no need of a reading light in those days and indeed haven't developed the need yet, but I did enjoy putting it on and off by pressing the little button on the end of the toggle that hung down above my head.

As well as the fine view it commanded of the farm close, with its sights of evening revelry and sounds of the morning's work, I loved Mary's Room for its wallpaper. It was one of those large patterned floral designs of a plant that was like a rose but had thick woody stems like a rhododendron. The plant grew endlessly up the wall and evenly round it with never a sign of reaching the thin branches near the top or the sparse ones at the sides of the plant. I used to put myself to sleep at night tracing a climb I would make along those stems and in the morning I would amuse myself until it was time to get up by climbing them again.

I never thought it odd that my room was called Mary's Room, though I might have had I known that my granny's name was Mary, for I knew where she slept. In fact it was called after the fourth of the Mackie children, whose room it had once been.

Mary had been the brightest of the three Mackie girls and by far the most beautiful. She had even been spared the infamous 'Yull logs' – the strong, rather featureless legs that were passed down to most of the family through her mother – Mary Yull. She had got her highers with flying colours and gone straight to university when she was just fifteen. That seemed wise when, still only sixteen, Mary Mackie won the first prize in Political Economy. But it turned out not to have been. The strain was too much for one so young and she had to take a year off to recover. However, she did recover, got her degree and a fine job with Marks and Spencer putting her political economy to work. I wasn't aware of it but when I was occupying her room Mary was being something of a black sheep in London. She had married a poor newspaperman and was living in conditions which her mother considered to be far from adequate. Journalists working on national daily papers were not normally hard-up but Basil Large, though a qualified surveyor, was a communist and worked for the party newspaper. The *Daily Worker* paid its staff as much as it could in the light of the needs of the cause rather than Fleet Street rates. I have no means of knowing which was less popular at North Ythsie, the economics of the situation or the politics, but there is no doubt things were past the worst by the time I arrived there. By 1943 the Germans had turned and attacked Russia. With Hitler and Stalin having signed a non-aggression pact the British Communist Party had regarded the war as one of aggression by the British against the workers of Russia. But now that the Germans had invaded Russia the Party had changed its mind. The war by the British was now a just war against fascism. Mary senior and Mary junior were now on the same side in the main political issue of the day.

That must have been quite a relief for the family, because while Mary was decrying the British war effort George, who gave his name to the middle room upstairs at the front of the house, was in the RAF preparing himself for the day when he would cover himself in glory flying bombers over Germany. George's room was the smallest in the house, which may be why his mother put the biggest of her six large children in it. George is still, in 2008 as he was then, within a whisker of six foot four and sturdy with it. My father, who once shared the three-quarter bed with George, said that George filled almost all of it. John Allan struggled all night to maintain a small bridgehead on the bed with the aid of his teeth and his toes.

When I first remember seeing George Mackie he was looking very smart and dashing in the uniform of an officer of the Royal Air Force. I'd seen him before in the rogues' gallery (that was what my grandfather called the wall of family photographs in his dining room): there George stood half a head above twenty or so bonny young men waiting to be decorated for their part in bombing Germany to hell. All but two of the others were killed in action, but George survived. In the photograph he had just had the Distinguished Service Order pinned on his ample chest by King George VI.

Now you would think that after flying bombers twenty-five times over Germany nothing would frighten a man. But on this leave George was to face a far more gruelling ordeal. He was to make a speech of acceptance at a lunch given in his honour by the Tarves Literary Society. The thought of addressing all those people who knew and admired his father was far more intimidating to the young George Mackie than the anonymous hazards of enemy ack-ack and fighters. Certainly, the Tarves Literary Society would have known all about the young giant it was honouring in celebration of his Distinguished Flying Cross. His big sister told me, though he denies it still, that, in the intimacy of his mother's company, the bomber navigator wept in anxiety.

It wasn't courage the young George Mackie lacked, for despite the odds of ten to one against survival he volunteered for a second tour – to fly another twenty-five missions over enemy-occupied Europe. He joined another group of frighteningly young men who cheerfully undertook the most daunting assignments with little hope of survival. When he looked back on it forty years later Uncle George, one of the very few navigators who made it to squadron-leader, was able to say, 'We just wrote ourselves off, you know, and set about having a whale of a time as long as we would last.' And despite those odds, George did last. He was said to have a charmed life. He came back with his plane all shot up but he did come back. His fellows thought it was lucky to fly with Mackie, and indeed the only time his plane didn't come back, George had been taken off operations. It had been deemed unwise to let him finish his third tour because it would be so bad for morale to lose a man of his seniority. The crew flew on without Mackie and on what should have been their third-last operation they were all killed.

The philosophy of fatalistic hedonism seems, at a distance of sixty-five years, to have been the best and perhaps the only way, short of fanaticism, to get through the ordeal for which these young men had volunteered their lives. But there was at least one who thought otherwise. Maitland Mackie would have standards maintained even in those awful times. After the ordeal of the lunch the young hero went back to his squadron and continued to live his life to the fullest, but shortly a letter came from Tarves. It was from his father: it had been brought to his attention by the bank manager that George, who had no other means of support than his pay from the RAF, had run up an overdraft at the Tarves branch of the North of Scotland Bank, in excess of £6. 'You will need to learn, George, to earn your money before you spend it.'

That letter came from a man who had recently built up a fortune in land on borrowed money. But that was different. That was investment. I can only think that Maitland Mackie hid from himself the extent of his son's jeopardy. He would not have allowed the possibility that George might not live to profit from lessons on thrift and financial propriety. Through most of his life Maitland Mackie feared death and was appalled by the sight of blood. He took great pains to avoid seeing it, though he well understood what happened to the animals on his farms when they left them. I am sure he had no real understanding that the young man had written off his future in the cause of his country.

The remaining bedroom on the third floor at North Ythsie was called Catherine's Room. It was the one most recently occupied by its namesake, for Catherine was the baby of the family. When I stayed at Northies she was not far away, at the University of Aberdeen, where she studied medicine and was into everything. She was preparing the teas at the university sports day when she was called, with some impatience, on the loudspeaker to attend the High Jump pit to take her turn. She nipped down to the pit, apologised for being late, tucked her skirt into her knickers and, without any such nonsense as a warm-up, won handsomely. Catherine had no idea that she had done anything out of the ordinary and was astonished when I told her in 1962 that she still held the ladies' high jump record for the Aberdeen University Sports. I think she holds it still, for the university no longer has a sports day.

Catherine's attractiveness, which was very considerable, came from inside her and its outlet was principally her face, which radiated fun and sympathy. Indeed, it is questionable whether anyone as interested as she was in the welfare of others should have become a doctor. At the British Exhibition in London in 1951, I was taken ill. I can still see Dr Mackie crying for me at the suggestion that they would take out my appendix. She persuaded the surgeons to wait and see if I really needed surgery and I swear that that was concern for her little nephew rather than any well-founded medical opinion. At any rate, I recovered from whatever it was. I still have my appendix, which has never given me the slightest bother.

I can't envision the young Catherine's face without the plump cheeks pushed up into rosy pouches by a great smile and the eyes appealing right out at me with humour and love. Men of all ages, and women too, found that irresistible. At any rate, during the war the youngest Miss Mackie was beaming at the world with all the optimism of the baby of a big family. Typical was her attitude to banks. Like her brother George, she took her father's example rather than his advice on banks, which she regarded as places from which to get money rather than to deposit it. The manager at the head office of the North of Scotland Bank in Aberdeen was astonished when the third-year medical student appeared in his office, introduced herself and said she wanted to open an account.

'And how much would you like to deposit?' he asked.

'Oh, no. I don't want to deposit anything. I may do some of that later but in the meantime I've come for an overdraft.'

What's more, she got it. And my Auntie Catherine used her overdraft well, indeed, with panache. The day soon came when she was in a position which would have caused a lesser person some embarrassment. Her overdraft had grown to a level which caused the bank manager to write to his client seeking some reduction in its amount. Having seen her father in and out of that disconcerting position all her days the medical student was nothing daunted. She wrote the banker a cheque (drawn on the same account) 'to be going on with' and, in the covering note she promised to let him have another small cheque 'to account' at the end of the month.

Catherine Mackie's grasp of economics cannot be said to have improved over the years. Fortunately, though, when she found the nearest thing she could to a clone of my father, he was a successful journalist on Fleet Street. And Ian Aitken was ideally suited to being Catherine's husband in another respect. Despite his degree from Oxford in Politics, Philosophy and

Economics, Ian hadn't enough interest in the nuts and bolts of money to be much worried by his wife's cavalier attitude to it. And he was employed by real newspapers like the *Express* and the *Guardian*, which paid him well. That meant that with Catherine's salary as a doctor they managed to scrape by – even in London.

My grandparents' other three children had left North Ythsie too long ago for their rooms still to bear their names, when I went there in 1943.

4
Jean, John and Mike

The Mackies' first-born was my mother, Jean, who was away doing her bit for the war effort as a teacher while I stayed with her parents. For reasons not obvious from her early photographs, they decided that Jean would be plain and made the mistake of telling her so. That proved a self-fulfilling prophecy, for my mother always went for warm clothes, sensible shoes without heels and shampoo designed only to get hair clean. She'd been to school at Miss Oliver's School for Young Ladies in Aberdeen. That was where the better-off country girls went, not so much for education but, as Ella Bell, a contemporary from near Fraserburgh, described it, 'a polish-up' in preparation for the tests of the wider world. Despite that, my mother acquired a taste for english literature, came first in all Scotland in her Highers and got a Master's degree from Aberdeen University. She taught drama to adults and Highers to children, worked as a journalist on the notoriously right wing *Daily Express* and redeemed that by getting an article not just in *Punch* but reprinted in *The Pick of Punch*, twice. And she married my father, who was at that time an officer and 'temporary gentleman', as he put it, in the Gordon Highlanders.

Jean was followed by two boys. John was farming at Bent in the Howe of the Mearns. He was the socialist (converted by Jean), became a junior minister in Harold Wilson's government, was chairman of the Forestry Commission and ended his working days as Lord John Mackie, Labour spokesman on agriculture in the House of Lords. Like so many old Labour socialists he accepted a seat in the Lords as the only way he could continue in active politics, but his wife was furious. She had been brought up plain Jeannie Milne from Kincardineshire and she was still the same person. In 2008 she is still going strong and woe betide anyone who tries 'Lady Mackie' on her. Jeannie's attitude was a particular disappointment to her daily help, who was banned from saying that she worked for Lady Mackie.

Then came Mike, who was farming at Westertown of Rothienorman. He was a Liberal and twice stood for hopeless seats in Aberdeenshire, but his main political focus was on local government and on education. Mike eventually became Lord Lieutenant of the County of Aberdeen, the first man with a peasant rather than an aristocratic background to hold the post.

Those two boys had been quite a handful. John in particular was always in trouble. It wasn't that he was a hellraiser as such. He was a serious child with a passionate interest in everything. He wanted to know how everything worked, he wanted to work it and he wanted to work it now. John's reluctance to take 'no' for an answer, or 'just because' for an explanation, got him into a lot of trouble . . . especially with Miss Thompson, the primary teacher at Tarves school. She must have been an awful woman. Certainly my mother still hated her seventy-five years after leaving her untender care. For Miss Thompson beat the little boy who went to school all bright-eyed and curious at the age of four – every day, or so it seemed to his sister. He was beaten for being late when there were interesting things to be investigated on the road, for answering back and for mistakes of the bowel or bladder, for discussing things in class and for asking questions interminably. When John handed Miss Thompson a bunch of flowers he'd picked for her on his way to school she said, 'you needn't think that'll save you', and gave him the strap for being late. His big sister never forgot the little round figure going out to face six of the strap with water running down his cheeks and his legs. She remembered the tears on her own cheeks, but she still felt the rage.

Rather wonderfully, Miss Thompson failed completely to dim the light in John Mackie's eyes. School had to be endured but that still left a lot of time for getting into everything and organising his brothers and sisters and the cottar bairns on the farm. No one who was on them ever forgot his hunting trips. The only weapon they had between them was the ancient pike that for some reason lay alongside the mangle in the basement of the farmhouse at North Ythsie. To that were added wooden spears and swords, and home-made bows and arrows. Extensive provisions were stolen from the neep shed, the garden and the kitchen, for John's hunts were planned to last a long time and cover a wide area. Despite that everybody always seemed to be home by suppertime, the provisions gone. And, even after they were told to take salt to put on the tails of the wild animals they would come across, they never caught a thing.

Mike, less than two years behind his big brother, shared John's interest in things but managed somehow with a little less abandon. That may have had something to do with the uncertain start made by the Mackies' third-born, for he had been a sickly child. He also had rather large and, in his father's eyes at least, sticking-out ears. So important did those ears seem to Old Northie (as my grandfather was sometimes known) that he appeared home from town one Friday bearing one of those caps rugby players use to protect their ears when scrummaging. My granny was outraged at the suggestion that the bairn should wear this in the hope that it would train his ears to lie in to his head. 'If that bairn lives I'll nae care about the set of his lugs,' she told him. Mike did live and, so far as I know, that was the last time the lugs were mentioned.

More cunning than his big brother, Mike survived Miss Thompson, without affection but without any real bother despite John's unfortunate precedent, and the two made life an odyssey of discovery and invention. There was a boat to be designed and built, and launched on mill dam at Little Ythsie. The Farmer of Littlies was so alarmed by this that he insisted Maitland Mackie get the sailors to desist. There was a cairtie to be built so that the two boys could hurtle down the steep brae towards Tarves. When they were making that they got a more or less round wheel off an old draff hurlie for the front but were quite stuck for a pair of wheels for the back of their chariot. However John, the chief designer and engineer, sent Mike up to Tarves to the saddler, who also did a bit of rather unprofessional cycle-mechanicking, to buy wheels . . . and he gave his procurement officer the following rather daunting guidelines: 'Noo, Mike,' he said gravely, 'if they cost onything ava, dinna buy them.'

Somehow the cairtie was built and the inventors moved on to greater things. It was only about 1920 and yet the two heroes set about designing the first aeroplane to be financed and built entirely in Formartine. You could be forgiven for thinking that such an undertaking would be beyond the means of an engineer whose budget was so severely constrained that he couldn't buy two second-hand wheels for a cairtie, but you'd be wrong. For Formartine's first aeroplane was to be built in the finest traditions of British invention, with whatever was already available. That ruled out an engine, and though manpower was considered it was eventually decided that North Ythsie One was to be powered by gravity.

The design that emerged after several hours of careful planning was an old tarpaulin ruck cover with ropes tied to the corners. By those the pilot was to be tied to the tarpaulin and for extra stability there were two more ropes onto which he could cling.

'Ah but,' I hear you say doubtfully, 'where on a farm in those days would you get a structure high enough from which to fly such an invention?' Well, on most Aberdeenshire farms there would have been nowhere very high in those days, but on top of Northies' hill stands the Prop of Ythsie . . . almost ninety feet high to the top of the battlements. Off that mighty monument the boys expected North Ythsie One to fly quite a way. Perhaps as far as Newburgh and the North Sea. On a fine day it doesn't look far to the sea from the top of the Prop, though in truth it is almost five miles as the crow flies. And there could be no guarantee that the aeroplane would fly as straight as that, not horizontally anyway.

The honourable job of test pilot fell to Mike. John was the engineer, George was clearly too young and the other three siblings were just quines. But as the three young men struggled the half mile from the factory to the launch site, the test pilot became less and less sure that it was a good idea. By the time they had made it to the top John was beside himself with exasperation. Could Mike not see that the ruck cover would spread out and catch the wind and let him float gently down and even along, as there was a good wind blowing towards the sea? And anyway if it didn't open properly Mike could spread it out with his arms.

They had started to tie Mike into the ruck cover when, as can happen, panic produced a stroke of genius. 'What about trying a steen first, John?'

'Oh for God's sake! Onything to please ye. Gwa and get a steen frae the dyke doon there on the Target Park.'

The stone proved to be a good deal more difficult to tie in than Mike would have been and that tried John's patience even further, for he could hardly wait to see the great bird fly.

Eventually all was ready. Six pairs of hands held the contraption over the edge and spread the cover as well as they could. 'Aye and ye see, Mike, when it's your turn it'll be easier for you'll be able to spread it oot yersel.'

At the count of three the boys let go. Without any discernible braking effect from the ruck cover, the stone hurtled to the ground. But fast as it fell, the stream of stone, rope and tarpaulin only just beat Mike to the ground. As soon as he saw the launch he was down the stairs and away through the parks and back to the woman who had been so much more concerned for his health than for the set of his lugs.

Mary Mackie, for all her generosity of spirit towards her young, couldn't understand why her husband didn't employ a more tangible discipline with his children. The Mackie boys led a considerable community of cottar bairns on a series of exploits along the rooftops of the steading, which meant that before every winter a high proportion of the windows and slates had to be renewed and, for all that money was often short, all that Maitland would do was to pay the bills and say that they had all been young once, though, when I was in his care, I could not believe that my grandfather had ever been as young as that.

Life wasn't all devilment with the young Mackies. As well as the usual helping round the farm running errands, picking and planting tatties and acting as gates when there were cattle to be herded, by standing with their arms open wide to indicate the required direction, the boys did what was reckoned to be useful work in discouraging the rooks. These were supposed to do great damage to the crops in those days when they were given no credit for the role they played in the battle against tory worms, the grubs which can play havoc with young cereal plants. The three lads had but one gun between them but that didn't mean they were two guns short. In the Mackies' understanding of things three was the ideal team. That was because they had somehow worked out that crows could only count up to two. If all three boys were seen going into the hide with a gun and two subsequently came out again carrying a stick, the crows thought that the boys had all gone and that the coast was clear. The innumerate crows were at the mercy of the marksman who was left behind. That was usually John. It was sometimes Mike, but it was never George, who was always 'the youngest'.

George complained to such an extent that the great day came when Mike agreed to give him a practice shot in the cornyard. Mike determined to make the experience as discouraging as possible. 'Noo mind, George, a big gun like this has an awfu kick so you should hudd it as far awa as ye can.' Instead of holding it firmly in to his shoulder as he should have, George held the double-barrelled twelve-bore shotgun at arm's length, pointed it at the tin can on the post, shut his eyes and pulled the trigger. Mike hadn't been exaggerating. Although a very big lad for ten, George landed flat on his back, but, perhaps through luck, with no injuries except to his pride.

John the Aviator's ambition at the Grammar School in Aberdeen was to play rugby for the first team. At the age of only fifteen he was on the cusp of fulfilling that ambition. It seems impossibly tough, but his father needed a grieve to run his outfarm of Little Ardo and a suitable man was not to hand. Then Maitland Mackie's long-time manager of Westertown, Mr Cumming, said to his employer, 'You have a very good man at the Grammar School.' So the fifteen-year-old boy was put in charge of what was then a big farm, with six men under him.

It was vital for the boy to get off to a good start, for those were good men and proud. If they were to respect this schoolboy as their boss he mustn't falter. The first impression would be made in the stable when, at six o'clock on the first morning, the boy would give the men their orders. Henry IV on the night before the Battle of Agincourt cannot have given more thought to his speech. John hardly slept a wink. The men would be lined up and he would walk down the line giving them their orders without hesitation. The first two horsemen would be put to the plough, the third would yoke his carts and bring in turnips, the orraman would pluck turnips and the haflin would take the orrabeast to Kelly sawmill for sticks.

At six o'clock they were all there except the baillie, who was among his cattle. The grieve, who was a full year younger than the haflin, made a start at breathless speed. The first horseman was told to yoke his plough and go to the sawmill for sticks. The third was told to yoke his cart and go with the foreman to the plough. The second was told to get his tapner and pu neeps, and the loon and the orraman were sent to bring them home. It was an inauspicious start, but the men knew what was required and went about it. Their only cruelty to their young grieve was to make sure that the whole countryside knew about it.

Everyone enjoyed the young man's discomfort but many said it was a shame that John didn't get more formal education. Certainly he had the curiosity and never stopped teaching himself. He read widely, elbows on the kitchen table or at a desk. He regarded reading in an armchair or a bed as too soft for the level of concentration to which he aspired. When I came to North Ythsie during the war John was thirty-five and farming on his own account at Bent, the finest farm in the Howe of the Mearns.

Mike too was farming on his own account, at Westertown of Rothienorman. He had got the education but it was quite a struggle to get him his BSc in Agriculture in 1932. He did his studying in the billiard room at North Ythsie and it had been the job of his big sister, my mother, to go up every quarter of an hour to wake him.

George was in the RAF, Mary and my mother were away working and Catherine was at the university. So I had my grandparents and their enormous, wonderful house and garden all to myself, when I was a refugee.

5

John, The Great Mackie Progenitor

The farm of North Ythsie covers the next hill to the east of Tarves, which is about two miles by road from the farmhouse and buildings. The steading and the house sit well into the south face of the little hill at the top of which the Prop stands monument to the farm's most famous laird. Sadly, and I do believe unfairly, his fame is greatest for being the prime minister who sent 50,000 left boots to the Crimea for the comfort of his freezing troops, in that terrible chapter in the Balkan Wars. His premiership came to an end in 1852 and it was some thirty years after that that my great-grandfather, John Mackie, took the tenancy from Lord Alexander Gordon's heirs. It was a mere outfarm, for John Mackie, whose home farm was at Auchnagatt, farmed also in Methlick and Rhynie.

For all the haste with which I entered this world I was too late to meet this John Mackie. To me he is just a series of anecdotes and a handsome, bearded, rather grim face looking down from the rogues' gallery, but those old photographers made everyone look serious. They made their subjects hold their poses for ages to give the camera a right look and you had to stay still or your movement would be recorded and make the picture look fuzzy. Outside photographs, grim, John Mackie was not. Indeed he was a most gregarious and charismatic man – and a man in a hurry. He had to get round those farms and the markets from Huntly to Maud and Aberdeen in the days when the gig was the fastest method of transport, and John Mackie's gig is said to have been the fastest in the land. When he came to see that all was well at North Ythsie, he would have his dinner and then jump into the gig which the strapper would have had ready and waiting and would be round the corner before he had finished asking for the strapper's family or giving him a last-minute order. Then he would be off to his hill farm, Milton of Noth in Rhynie. That was a day's drive to anyone else but John Mackie would be there for lowsin time at six. Word would have gone round the foothills of The Tap o' Noth that John Mackie was coming and by early evening his pals would be gathering at Milton of Noth to share his whisky, his conversation and his billiard table, of which he had one at each of his farms.

It had been a particularly difficult lambing in 1903 when, on what was the spring day elsewhere but still winter on the north-lying slopes of Tap o'Noth, my great-grandfather arrived at Milton. The 300 ewes had nearly all lambed at night. The lambs were none too strong, for it had been a bad year for hay and a long hard winter. Alec Thomson, the shepherd, had done a wonderful job. For four weeks he had dedicated himself to his ewes and lambs, snatching only the odd wink of sleep here and there between crises. So when the farmer arrived in his gig the shepherd was just about done in, though, good man that he was, he was hurrying home for a quick bite before another night's vigil. John Mackie summed up the situation in an instant and ordered his man home to bed with the assurance that he would do the nightshift himself. And he didn't want to see the shepherd about the place until morning.

After his supper the old man checked the sheep and found, as he expected, that the shepherd had done a first-class job against all the odds. He went through the byres and stables and saw that all was well with the beasts. There were plenty of neeps in and corn thrashed and bruised and there was still enough in the cornyard for another two days' thrashing. That would be one for the horse and one to sell. Yes, John Mackie was well pleased with how things were going on the farm.

The housekeeper had laid in plenty of whisky and cheese and had done a baking of oatcakes and dumpling. With five of his best pals, the flower of Rhynie's middle age, he set out on a night of conversation and billiards. The old man had been a great conversationalist and by all accounts he was a wizard at billiards. I can well believe it, for his son Maitland could play a shot I still find remarkable. By using top spin, not the side and bottom those television snooker stars use, he could make the cue ball follow the other into the same pocket even when they were well off line.

But I digress. On that night in 1903 the conversation and the billiards and the roaring fire and the cakes and ale had proved of more immediate importance to the old man than the ewes. Suddenly it was morning. 'Oh my God, the sheep!' He and the remaining two guests hurried round to the buchts lest Alec Thomson beat them to it. Luckily there had only been one lambing. The ewe had one fine sturdy lamb up and going but unluckily it also had one newly born and lying still. What a disaster! Not the loss of one lamb, you understand, but the potential loss of face at having failed to keep his side of the bargain, after all the shepherd's hard work over four weeks. There is no doubt that the shepherd, exhausted as he was, would rather have watched his sheep than abandon them to such neglect. There was only one thing to do. They would bury the lamb and the shepherd would never be any the wiser. But where to bury it? To that the conspirators gave too little thought. They decided the muck midden would be best as it was handy and an easy place to dig.

When Alec Thompson appeared at seven it was, 'Aye, Shepherd. I lookit the ewes last nicht – twice. There was jist one, a single.'

'A single was it,' said the shepherd in a fury. 'Well what the hell's this? I found it rinning aboot in the midden bawlin for its mither.' With that he produced a fine perky lamb from inside his jacket. The warmth of the muck had been all the lamb needed, and the old man's laziness had meant he wasn't too deep in it to escape.

I feel that that story should have a moral though it is usually told without. This will do. It is one of the wisdoms of James Low, the grieve at Little Ardo, of whom more later: 'If you're gaun tae tell a lie, tell a good een,' he used to say, and there's no doubt he knew.

More of John Mackie can be understood from the time he took sixty-four country children from Auchnagatt to the great Barnum and Bailey's Circus in Aberdeen. Now, in 1890, even though it was after the coming of the railway, that was a journey of some magnitude. It was nearly twenty-five miles and the idea of taking the entire Sunday school was considered daring to the point of foolhardiness.

'You're bound to lose some of them, John Mackie, and how will you feed all that bairns, and what if you miss the train and you'll never keep track of them all in Aberdeen,' they said, and worse.

'I'll easy manage,' said my great-grandfather, 'never you fear. I'll count them all onto the train and I'll count them back off it again at Aberdeen. I'll count them into the circus and I'll count them out again. I'll count them back onto the train and YOU can count them when we get back.'

The total cost of the great day would be no less than four shillings a head despite the old man's efforts in getting a discount from Messrs Barnum and Bailey and the excursion rate on the train. Now, that might not seem a lot today, but to the cottar families of Buchan in 1890, with their high fertility and low wages, it was far beyond what they had ever contemplated spending on a treat for their children. But there had been a small surplus on the Highland Games which were run in aid of the hall fund and John Mackie was able to see that some money was made available from there. In the end, a nominal charge of one shilling was to be made. Even that was too much for some, but private arrangements were made so that no bairn would be left out.

When the day came, every child in the parish of Auchnagatt between the ages of five and thirteen assembled in front of the Baron's Hotel. They were in their Sunday best – which in many cases wasn't that good – clean, cropped and eager.

They had been brought by their parents, some of whom wept, for this would be the first time there had been anything like twenty-five miles between them and their children. There were those who were unkind enough to say they were quite right to weep, for John Mackie would never manage to bring all sixty-four through such an adventure.

Six children to each of eight gigs and the rest in a phaeton (a four-wheeled carriage), the gallant sixty-four set off to a breathless cheer and swayed and bumped along the six rough miles to Arnage Station.

There John Mackie duly counted his charges onto the train; ten to a carriage and the smallest fourteen with him. Each had a bottle of hairst ale and a quarter of oatcakes. The train took off to another fervent cheer. Excitement was high. Few of the children had been on a train before and some hadn't been outside the parish of Auchnagatt. The youth of the parish were experiencing an expansion of their world at once sudden and great.

When they reached Ellon, a good five miles away, John Mackie leapt out of the train and ran along the platform for another count of the troops. It was then that he met the first crisis of the day and one for which he had not planned. What with all the excitement and the hairst ale, and there being no corridors on the train and no facilities, half the children were 'needin' and some were 'awfu needin'.

What to do? If he let some off, how could he be sure he had them all back in their tens? And there was little doubt that as many more would be 'needin' when they got to Udny Station and as many again at Logierieve. There was only one thing for it. He would get them all out at Ellon, 'needin or no'.

Now the facilities at Ellon were hardly made to take sixty-four at a time, but country bairns were resourceful in those days and John Mackie soon had them counted back into their compartments and, with only a slight delay, the guard waved his green flag, blew his whistle and, to the obvious relief of the station master, the train pulled out.

At Aberdeen's Kittybrewster Station the sixty-four were counted off the train and herded the half-mile to the great Barnum and Bailey's marquee, which had been pitched in Central Park. I won't try to describe the wonder and excitement of those country bairns at seeing for the first time a town, a big top, lions, clowns and elephants, but by all accounts they were intense. One little boy said to his mother when he got home, 'Oh Mam, and there was a pooler bear' – which must have been an exaggeration.

After the show they were counted out of the marquee and counted again when they sat down in the Kittybrewster Tearoom. They were still all there for their white fish, tea, pieces and jam . . . and cakes.

I feel unable to do justice to the great day by telling you of all the times that someone got lost or someone else thought they were lost, or someone else hid or stole someone else's piece, and in any case I'm sure you can imagine it all. At any rate after many crises and much hassle the train carrying the country bairns home pulled with a sigh of relief back into Arnage Station. From there the convoy of gigs and the phaeton carried them back to Auchnagatt where their anxious parents were waiting. Would Mr Mackie have brought them all home? And there were some who had warned of the foolishness of the venture who were hoping that he would be one short or even two.

As the convoy hove into sight and wound its way down the hill to the Baron's Hotel, the counting started. They weren't easy to count but there did seem to be plenty of bairns. 'Good God! He's got sixty-five bairns. He's got an extra een. Oh dear me! John Mackie's fairly deen it noo.'

Consternation, along with the delight of scandal, reigned. What would Auchnagatt with its legendary fertility do with an extra bairn? But when the convoy had disembarked and the anxious parents had each claimed their own, there wasn't one left over – each had got a home.

It was quite a mystery and John Mackie waited years before telling the village that the extra bairn had been at the dentist in Aberdeen and, rather than miss the fun, had arranged to meet the others at the circus and get home with them.

No one really knows what John Mackie was playing at, for he never told anyone. Perhaps he was just trying to empty a stuffed shirt. But it is one of the favourite stories in my family, where stories are much cherished.

John Mackie was one of the first farmers in the North-east to get a reaper-binder. This was a really great innovation. Prior to this invention the corn had been cut by a reaper, a horsedrawn mower, which left it lying on the ground in lines which were known as bouts. The bandsmen, and often bandswomen, then had to gather the bouts into armfuls, make a band from a handful of the same corn, and then bind the armful together to form a sheaf. Those sheaves were then propped up in groups of eight, sometimes ten, called stooks, where they were left for however long it took to get them ripe and dry.

There was still plenty of work left in the hairst park, but at least the binder did away with the bandsmen. The reaper-binder cut the sheaves and bound them automatically with binder twine. On those farms where the work was mostly done with a shovel, a graip and a barrow, the binder was a pretty sophisticated piece of kit. If it broke down or jammed there was not, in the early days, anyone with much experience of how to put it right. And so it was that my great-grandfather gained a quite unmerited reputation as some sort of mechanical genius. He came upon the binder stopped and the boys all scratching their heads and wondering what to do. John Mackie joined in the head-scratching. He was quite upset. This machine was supposed to speed up the delicate job of harvesting. If it wasn't going to work they would be much better without it. At least they could revert to the old reaper and get the bandsmen going again and he could have saved the expense of buying the damned thing.

In a fury of frustration John Mackie lashed out with a kick at his new machine. Now, as he told the story, there had been a small stone jamming up one of the many chains that transferred the power from the wheels of the binder to the mowers, conveying cloths and tying knives of the binder. This great boot dislodged the stone and the binder was restored to full production. There were those who swore that Mackie had known exactly what he was doing when he kicked the machine – but he wasn't one of them.

John Mackie was widowed early. His first wife, Jean, of whom nothing touched me save for one stunningly beautiful photograph in the rogues' gallery, died after bearing him a daughter to add to their four sons. The doctor who attended her had been at a case of scarlet fever and brought it with him, killing mother and child. John Mackie was twenty years a widower before he took a second wife, Elizabeth. Then his son John took over the home farm of Mains of Elrick, Willie took over at Skelmafilly in Methlick, Tom bought himself a farm, Broombarns, in Angus, and Maitland moved from grieve at Rhynie to be farmer at North Ythsie of Tarves. Old John Mackie moved to the farm at Rhynie and made a fresh start. Elizabeth bore him two children, Bruce, who followed him on the farm, and Elizabeth, who married into the Stephen family at Conglass in Inverurie and so established a connection with Virginia Woolf (née Stephen), of which some in the family are still proud. John Mackie died in 1914 in his sixty-third year, and I remain sorry that I missed him.

6
Maitland and Mary – My Grandparents

It was their six children who earned for the family the reputation 'the Mackies are like whales – they maun come up tae blaw'. Their male children certainly did make an awful lot of noise. They were huge for a start, they had big voices, big ideas, unlimited self-confidence and seemingly unlimited capacity, not just for food and strong drink, but for getting things done. They were always in the thick of it, whatever it was. No issue was too local nor too wide-ranging for the Mackie boys to get into and they were always in the centre of it. Yet no one would ever have said that their parents were ones to blaw.

Maitland and Mary were married at her family farm of Little Ardo of Methlick in 1906 when he was twenty-one and she was twenty-three, and set up house at North Ythsie, from which Maitland built up an empire of farms sufficient to settle each of their six children in at least one substantial farm of their own. They were married for sixty-five years, all of which were spent together in the same house and, apart from two occasions when Granny had accidents and had to go into hospital, in all that time they were never apart for longer than a day.

There must have been some hectic times when the six very strong and strong-willed children were growing up and the out-farms, which were up to fifty miles away, were being bought or rented and farmed against the background of the Depression between the two world wars. But when I went into their care during the war and for the twenty-five years more which they survived there together, Maitland and Mary Mackie created at North Ythsie an atmosphere of serenity, humour and above all, order. Maitland was always sitting on the right-hand side of the fire on the telephone, reading or playing patience. Mary, if not in the kitchen or the garden, was always on the left-hand side of the fire and usually knitting. Half of Aberdeenshire seemed to know that you'd get coffee at North Ythsie at ten in the morning and afternoon tea at three. It was a bit like a hotel and when I was there, I loved it.

Considering how handsome both John Mackie, his father, and John Mackie, his son, were, Maitland wasn't a particularly good-looking man. His ears and his nose were too big for classic good looks, but at six foot three, and never more than sixteen stone, he was a splendid figure of a man who held himself up and looked eagerly at the world from beneath a selection of felt hats. Alec Watson, his tailor in Aberdeen, was able to make suits for Maitland Mackie for the best part of fifty years without a fitting.

Mary was also tall for her times at five foot six inches, and erect even in her eighties. A good-looking rather than beautiful woman, she dressed well but not flashily, though she did like a silk scarf held in place by a pretty brooch. They made a magnificent couple, and it is a great loss that there isn't a photograph which adequately catches that. For some reason Mary put on a rather prim semi-smile and stood rather stiffly for photographers, relaxing into her gorgeous laugh only after the photograph had been taken. Mary Yull's one physical flaw was her thick legs. They certainly didn't hold her back, for she was a beautiful mover who achieved glamour way beyond anything that she sought. But those legs, passed on even unto the third generation, meant that the men tended to look well in the kilt and the women in long skirts.

Apart from with cameramen, I never saw my grandmother in a situation of which she did not have complete control. It gave the young refugee such a feeling of safety to be with her . . . there was no fear of not knowing what to do or doing wrong . . . everything was certain and there was no need for foxes. No doubt all that certainty was something to do with Mary's faith in a Christian god. She tried never to miss church. She did as little as possible on a Sunday and she had a habit which many found disconcerting and her daughters thought outrageous, of asking guests, even complete strangers, without warning, to say grace at the table. She wasn't bigoted in the sense that she was against the Free Church, Episcopalians or Catholics (though the last didn't arise in Aberdeenshire in her day). As long as you believed in a Christian god you were all right, but my granny was sure that communism could never succeed and for that difficult judgement all she needed was that it was 'so unChristian'.

Mary did good works, of course. During the war she had a little round of the poor and the lonely to whom we brought a pottie of jam maybe or a knitted thing, and our company. I used to enjoy our visits to old Mrs Fife, the silver-haired widow who lived in the little croft beside the Sonah Bridge where my father and the minister had written their names in the snow. I was astonished to hear, years later, my granny remark on what an angry person Mrs Fife had been. To me she'll always be warm and sweet. She always had a pan drop for me.

My grandmother's Christianity was heavily influenced by the Puritans. In 1943 her house was teetotal – my granny had been put off the drink by the excesses she had witnessed in her parents' house as a girl. In particular she was infuriated by the deterioration she saw in a neighbour who had been a bonny young man, full of wit and fun, who was reduced to the point where he could make his water against the table-leg in the kitchen at Little Ardo. There had been some drinking at North Ythsie in the early years of their marriage but Granny put a stop to it. According to Mrs Shepherd, of the neighbouring farm of Shethin, Granny had called a halt after she had had to send the grieve over to Shethies to get Maitland home at six in the morning, when it was time for the men to yoke their horses. Shethin and Old Northie had been great pals and my granny suspected they had been over-indulging at the mart, when they came inbye and the taciturn Mr Shepherd had been very chatty while Maitland, who was normally full of news, was uncharacteristically silent.

And Mrs Mackie didn't care for excess in eating either. She disliked the excellent feasts put on by her daughters-in-law and decried them as 'vulgar'. At her table there was always a beautifully cooked and presented elegant sufficiency and no more. It was one of Maitland's proud boasts that he could put a boiling hen round a dozen people and keep them all laughing throughout, and I do believe that, more than once, he had to. So much did Mary frown on gluttony that when her three large sons were invited to a party, she would not let them go until they had eaten a huge plateful of porridge, lest at the table they disgrace her.

My favourite memory of Mrs Mackie of North Ythsie is of a fine sunny day and she in her beautiful and productive garden picking pinkies for sale in the shop in Aberdeen. I loved her then. But I don't really think she loved me much. She just produced everything that made it easy for me to be happy. The Mackies had the ability to make everyone they met feel special and I fell for it completely. I was secure in my knowledge that I was their favourite until some twenty years later. On two separate occasions I heard two of my eighteen cousins on that side who were not only pretty enough to be girls but actually were girls, say 'of course, I was their favourite'. There being nineteen grandchildren, Mary probably didn't make much distinction between us but those girls still believe they were Granny's favourite, whereas I know I was.

She wasn't a saint though. Mary was unable to forgive her illegitimate sister Isie, who had been taken into the family, for her father's misdemeanour in having a child out of wedlock. Her certainty sometimes made her intolerant of weakness in others – like the time she wouldn't stop complaining about the cottar wife who she was sure was stealing eggs. Maitland put a stop to that with one of his only recorded reproofs to any of his family, 'Hudd your tongue woman,' he said, 'if you'd ten bairns in a cottar hoose you'd steal eggs yoursel.'

Maitland Mackie's success as a farmer was remarkable. From a 400-acre tenanted farm on the second-class lands of Formartine, he built up an empire of 3,000 acres of owner-occupied farms stretching from Buchan to the Howe of Strathmore in Angus and including 1,500 acres of the best land in Scotland. He founded Mackie's Aberdeen Dairy Company, with its shop in Aberdeen, and still found time to lead the National Farmers Union of Scotland, direct Lawsons of Dyce, the enormously successful pig products company, and govern the North of Scotland College of Agriculture. He also helped a large number of farming friends to run their farms. Those included James Keith of Pitmedden House: when Keith went off to the First World War, Maitland Mackie looked after his thousands of acres. Mind you, Major Keith didn't saddle his old friend with the job a moment longer than was necessary. The day the armistice was signed, on the eleventh hour of the eleventh day of the eleventh month in 1918, he thanked and said goodbye to his troops and set off for home without waiting for any such niceties as demobilisation. Those were, in some respects, the days.

So how did Maitland Mackie manage it all? Well, I certainly can't claim to understand the alchemy of success in farming; if I did I would have made a better job of it myself. But I can recognise some elements of his success.

The first, and I have to be very careful about this, for my grandfather was a generous man, was that he was very keen on profit. He would never give up on an idea for fear of losses. But the prospect of progress was what motivated his farming and, in an imperfect world, profit was the best indicator of progress. He gave a lot of his time to public work and never sought personal aggrandizement from it. When he qualified for the pension he called the minister down to North Ythsie and told him he would like to give his pension to the church. He was promoting and raising the funds for a youth hall to be built beside the church. His and Mary's pensions could go towards that. The minister's reaction was greeted with some resentment among the family. The minister said, 'Oh yes. I suppose that would be all right.' The young Mackies, by which I mean my uncles and aunts who were brought up in the early years of the last century, thought the minister might have made a better show of gratitude, or at least enthusiasm. Perhaps the minister was thinking about camels getting through the eyes of needles and rich men getting into heaven, but I am certain Maitland Mackie wasn't. It is only as I write this, some sixty years later, that it has occurred to me that the old man must have been a bit hurt himself, for how else would that story, which would not have had any witnesses, ever have got out?

There is a family anecdote which makes a point about Maitland Mackie's priorities very well. It contrasts Maitland's attitude to some beautiful Morayshire woods, with that of his less serious and jollier older brother John. John had been the chairman of the North of Scotland College of Agriculture's governors and had been driving with the secretary of the college through the woods. 'These are grand woods, Mr Mackie, don't you think?' 'Oh yes,' said John Mackie. 'Great woods. A perfect place for a walk with a young woman on a summer's evening.'

Ten years later the secretary was still in post but had a new chairman; Maitland Mackie had followed his brother. The secretary found himself again driving through the same woods in Morayshire and said to the new chairman what he had said to his brother ten years before. 'These are grand woods, Mr Mackie, don't you think?' My grandfather took a minute to think and said, 'Yes, indeed. Good, well-grown timber. These woods could be worth £1,000 an acre.'

Though Maitland and Mary felt they were adequately provided for without their pensions, Maitland was much more interested in the pennies than in the sort of artistic tastes his daughters had acquired at their universities. Life at the University of Auchnagatt had not prepared him well for high culture. His daughter Jean, my mother, redecorated his traditional drawing room in pastel pinks and blues, and made up exotic colours no one could put a name to. The centrepiece was a chandelier made out of a highly coloured lady's umbrella suspended upside-down below the central light. Everyone agreed that something needed to be done about the naked 100-watt bulb, but an upside-down umbrella! She then festooned the walls with copies of the Old Masters. Old Maitland Mackie was delighted and most amused. He played a game with his educated daughter, swearing that the eyes from the phoney Mona Lisa followed him round the room.

He knew his job thoroughly, having been brought up on the farm by a very good farmer, and he had been removed from school when he was sixteen to work as an orraman on his father's home farm of Mains of Elrick. There Maitland had a hard apprenticeship. They say there were times when he wasn't fit for it. He was very big for his age, taller than any other of the men, and so was expected to do work that was heavier than that for which his young muscles were ready. On the first day of his first harvest he was gathering corn into sheaves behind the scythe of the second horseman. That was back-breaking work which involved being bent double all day. The scyther wasn't going to let the boss's son away with anything and kept up a good pace. It is said that Maitland was struggling when his father came past. John Mackie summed the situation up quickly. 'Aye Maitland, gwa doon the Howe. There's a bit o palin doon and if it's nae sorted the stots'll be among the corn.' The young man left the gathering gladly but, though he walked the length of the Howe, he couldn't find a break in the fence. He stayed a while not to give the show away and then returned to the hairst park.

That kind of understanding helped my grandfather through, and as the days passed his arms grew stronger and the skin on his hands grew thicker till he began to think he was as good as a man. One of many false dawns came after a potato gathering, when he and the other orraman were put to cover the tattie pit with earth against the frost. Side a piece and throw for throw, the boy found that he was able to cover his side of the pit as quickly as the man. The man said nothing but as spring approached the two were sent to uncover the tubers. The man made sure that he took the opposite side of the pit so that he was throwing off the earth that Maitland Mackie had thrown on and vice versa . . . then my grandfather saw how far he still had to go to be a real man. He did of course make it, and it was certainly earlier than most, but Maitland took off his tackety boots for the last time when he was thirty-four and concentrated on what he did best, which was organising the toil of others.

There the philosophy seems to have been to hire good men, give them the tools and then let them get on with it, with enough attention not to feel neglected but not so much supervision as to make them feel crowded. It was a formula that produced devotion in those most undemonstrative of men, the Aberdeenshire farm servants. He even had a grieve, called William Cumming, who put through the farm accounts the backhanders, those little tips a farmer or his agent may get when he buys livestock at the mart.

Another pillar of Maitland Mackie's philosophy was that the most important decision a man had to make was who to take as a wife. If a man made a good job of that he would likely make a good job of everything else. So when he went to fee (engage, or employ) a man he always tried to catch him in the house, perhaps at dinner time, so that he could see how that was kept. If the house was well scrubbed and the bairns well patched and darned then the chances were he would get a good man.

He always paid half a crown in the half-year above the going rate and that helped him to get good men. But it wasn't just the half-crown that caused the great horseman and grieve James Low to fee at North Ythsie. They met at Ellon feeing market where men and masters met to negotiate terms for the year-long contracts for married men and six-month contracts for single men. The half-yearly markets were the only times when the bosses and the workers met on terms of anything like equality, and the stronger men, and some that were only more foolish, took the chance to give the farmers a bit of cheek. And so it was when Maitland Mackie approached James Low at Ellon Market in 1927 and asked him to fee to North Ythsie. Low had been at a rally in the Town Hall at which Joe Duncan, the legendary founder of the Scottish Farm Servants' Union, had told the men that if they all stuck out for £32 for the half-year, they would get it. The men had then marched behind the Ellon pipe band, down to the market to tackle the farmers.

So with all that solidarity and a couple of drams inside him James Low wasn't going to let his eight-inch height disadvantage hold him back in negotiation. 'Ye hinna an awfu good reputation,' the applicant for the post of second horseman told the farmer of North Ythsie.

'Oh, aye?'

'Aye, yer men are ay gaun aboot wi a sack o tatties on their back,' said James Low.

Maitland Mackie did not argue. 'You might think more of me if once you were home,' he said.

And so he did, for James Low developed a regard for Maitland Mackie which, in societies where that word is used more freely, would be described as love and was certainly the very highest regard of which the rather dour people of the North-east corner of Scotland are capable. Whatever you call it, that regard kept James Low in the service of my family for forty-six years, both at North Ythsie and at our farm of Little Ardo. When he left Little Ardo he tried to buy a house that was once part of the farm, in the village of Methlick, because 'you wouldna be far fae yer hame'.

Mind you, not all Maitland Mackie's dealings at Ellon Market were that successful on that day. Having given one man the arles of half a crown ($12\frac{1}{2}$ pence) that sealed their bargain for a half-year, my grandfather, without any malice or even much thought, offered the parting words, 'Man, ye've tremendous feet.' Two days later a letter arrived at North Ythsie. It contained the half-crown and a note saying, 'If the feet doesn't suit the man won't either.'

The day came when James Low feared he might have been better to return his own arles. By this time he had made rapid promotion and been packed off to be grieve at Little Ardo of Methlick, seven miles away, with the words 'be hard but fair and never get too friendly with your men'. But Low was hardly started in his new job when a henhouse burnt down, with the loss of two thousand pullets – a huge number in those days, when the typical henhouse held about twenty birds. James waited the coming of his boss in shame, not knowing whether to leave or wait to take the sack face-to-face.

But when Maitland Mackie arrived he had brought a builder with him.

He asked nothing about the fire. He wanted only to get the rebuilding going as soon as possible and to know if his grieve had any suggestions that would make the new poultry shed better than its predecessor. The old man had a gesture with his hands which meant that he considered all was well and just as it should be. It consisted of spreading his long and useful but elegant hands out, palms up, and giving them a little tilt, rather in the manner that some prelates use as a two-handed salute. When the meeting was over and plans had been made for the new shed, James Low was glad to see that gesture; it may have been on that day that James Low's regard for Maitland Mackie changed from admiration to something stronger.

My grandfather was unflappable. He kept his eye on the ball. Like the time the steading at North Ythsie caught fire. The blaze could be seen for miles around and his friends and neighbours hurried to the rescue. They got the cattle out but the pails they had brought were useless against the flames, which had a firm hold. The fire was beyond control. Maitland Mackie didn't curse and swear and stamp his feet, or call down the wrath of God upon whoever had caused the blaze. He invited everybody into the house for an evening of tea and billiards, and left a hopeless case to take its course.

Old Northie was, as I've said, a progressive farmer. He was often into new ventures and was often the one who got his fingers burnt. He not only went into dairying (he had a dairy on each of his farms), but when he established Mackie's Aberdeen Dairy Company he also became a considerable retailer of 'Mackie's Milk'. It was his son Mike who made up the North-east's first advertising jingle:

Look at me sae big and strong
That eence was thin and wiry.
If ye want tae look like me
Tak yer milk frae Mackie's dairy.

The dairy is still making an impact in 2008. It was the seed from which Mackie's Ice Cream has been developed into a national brand, with international ambitions, by Mike's branch of the family.

Maitland Mackie can also claim a first in sports sponsorship, for in 1935, when the Bells team from Tyrie were recognised as champions, a Tug-o-War team was assembled from three of the Mackie farms. They trained at North

Ythsie after their hard days' work (and in James Low's case after cycling the eight miles from Little Ardo) and at the Great Turra Show they won under the very commercial name of 'Mackie's Milk'.

As with so much else, old Maitland was one of the first into barley beef, Landrace pigs, Friesian cows, disease-free pigs, deep-litter hens and battery hens. Those were kept at Woodlands, the Udny farm which eventually became Dr Catherine's, and I used to go with him on his weekly visit to see that all was well. The highlight of those visits was the incubator house, where tray upon tray of eggs were hatched in an electric oven. He had an incubator for 1,600 eggs, so there were sufficient that there would always be at least one tray with little fluffy balls cheeping their first and other eggs with sharp little beaks pecking their way into a life which he had ordained would be narrow and short.

The old man's many enterprises were always planned and counted on used envelopes. I used to wonder, as a very small boy, if this rather down-market vehicle for cash-flow projection was the reason the old man almost always lost money.

At the end of harvest, or when he sold his stots, the old man would be concentrating hard with his retracting pencil and an old envelope. Then he would emerge with, 'That's another £320 I've lost.' And yet the old man's business clearly sailed through every crisis, going from strength to strength. The young refugee used to worry sometimes where the next morning's bacon rind would come from. But I needn't have worried. I discovered much later that Maitland Mackie had his own way of making up a profit-and-loss account. If the envelope he had used to estimate the viability of a project suggested he should make £1,000 but he actually made £1,500 then that was indeed a profit. But only of £500. If the surplus was only £500, only half the figure he had expected, Maitland Mackie didn't just regard that as a disappointment – to him that was a loss. With accounting like that it was fairly easy to prosper through a series of 'losses'.

One Friday in the winter of 1924 my grandparents set off as usual for the markets and shops of Aberdeen. There were those farmers who still went by gig but the Mackies had a car. It was a 'Model T', and not just your standard model. Henry Ford had built the engine and the chassis in Detroit but the comfort of the car had been greatly enhanced by T. C. Smith, Coachbuilders, in Aberdeen. In their long lives, which included sixty-five years together, it was the only time the Mackies came near to disaster. It turned out to be one of the wildest nights in the history of the North-east. The day had started like any other Friday. Mary had done her shopping on Union Street and Maitland had been at the mart and seen several of his suppliers. The car had been at the garage for some minor repairs and when they went to pick it up at five o'clock the mechanic warned them not to try to drive home, because of the weather. But those were the early days of the motor car and there had been little experience of getting stuck in snow. There was hardly anything that would stop a gig drawn by a good horse and with a determined driver. For one thing, hooves are much more efficient than tyres in snow and for another, an open gig didn't encourage travellers to set out on wild nights. There is no doubt the warmth of the Model T Ford was partly responsible for the rash decision to set out on the fifteen-mile drive home on so stormy a night.

Maitland and Mary didn't find out until the next day that an urgent message to stay put had reached the garage just after they had left. It had been sent by the last person to make it back to Tarves that night. So they set off. As soon as they had cleared the town it became apparent that the journey would be difficult. It always does get worse the farther you get from the town and from the stabilising influence of the sea. The wreaths of snow were building up all along the turnpike. The Model T was fit for them. But what was worst was the visibility, for the wind was whipping the snow across the face of the car like a white sheet. The windscreen wipers were hardly strong enough and the lights only changed the white-out to dazzling gold.

After six miles or so Maitland noticed that, while he could see very little through the windscreen, and that only intermittently, if he opened his window he could see the dykeside fairly clearly. So driver and passenger each opened their window and they proceeded on the basis of the driver keeping clear of his dyke and the passenger advising him whenever he was about to hit her dyke. 'Over a bit Maitland,' she would say and he would swing to the right until he saw the right-hand dyke looming. Under that unlikely regime the two made steady, if slow, progress, until my grandmother mistook a snow drift for her dyke and insisted that my grandfather steer to the right until he landed in the ditch. There was no way out. The Model T was on her side and she would be there until the morning. Now, Mary and

Maitland knew quite well that the orthodox thing to do was to stay in their car and wait for morning or for the storm to abate, but they were only a mile from Maitland's brother John's farm of Coullie at Udny, so they decided to walk.

Their clothes were thick and well-made, but they weren't intended for those Arctic conditions. The gales whipped the snow into a powdery torrent which blotted out the road, stung their faces and choked their lungs. The drifting was filling the roads and they had to plunge through drifts that grew in number and size. They became really alarmed when they came across a drift in which, as they plunged through it, they found a fence. It was obvious to them that they were no longer on the road but in a field. They found a road again and staggered on towards a warm fire and hot food, but it was into the teeth of the gale and Mary was failing. Maitland tried taking her inside his greatcoat and reversing along the road but that was too slow. As they hit each new drift they fell down in the snow.

Eventually my grandmother could go no farther. So the old man decided to leave her and go the last half-mile alone and bring help. He took off his greatcoat, wrapped it round her and laid her down behind an enormous drift. 'Now Mother, you lie there in the lythe of that drift and I'll go to Coullie for help.'

He had almost gone when she called out, 'Maitland, Maitland! It's not a wreath. It's a house.'

And so it was. It was a cottar house on a farm next to the one to which they were heading. There can be few occasions on which a farmer was gladder to be cottared. In the morning my grandfather walked the last five miles home to North Ythsie. There he learned that his brother William had also been caught in the storm but had stayed in his car, where he had been very cold but relatively safe.

On that day Maitland Mackie showed that he had a temper to lose. It is the only time of which I have heard. His elder sons had been fighting, as was not unusual, when he arrived. He caught each of them a good thud on the side of head and said they should think shame to be fighting about nothing and their father and mother almost lost in the snow. The boys were somewhat aggrieved by their unusual treatment. They had been fighting about something of great importance to them at the time and had all along thought their parents had stayed the night in Aberdeen and were quite safe.

However, things improved for the boys, as the day developed into a great adventure. Mr Mackie yoked four of the five pairs of Clydesdales to one cart and set off to bring his wife home. The sixteen pairs of hooves battered the snow flat, making way for the cart. The sun shone on the dazzling wilderness. And when the boys met the cart they were allowed to ride a pair of Clydesdales each on the last mile home.

Maitland was a first-class games player: captain of the Auchnagatt cricket club as a young man (I can well remember his intense disappointment when, at the age of seventy, he was asked to stand down from the North of Scotland College of Agriculture's staff team); a demon of slice at table tennis; and a first-rate bridge player, held back by his partner's tendency to overbid her hand. But my grandfather's best game was chess, at which he took on succeeding generations of his family and beat them all until age started to go for his concentration.

It was a very important ambition of mine for many years to be the first of his descendants to beat Maitland Mackie at chess, but he held on grimly, leaving no act of gamesmanship unplayed. The first time I got into a winning position, he kept playing, more and more slowly, and refused to resign until it was time for the bus that would take me from the end of the road back to Methlick and Little Ardo. A few months later I again came close and again bus time loomed with no resignation forthcoming. This time I was prepared. I went for the pegboard in the drawer of the display cabinet so I could capture all the positions and we could resume on my next visit. But when I came back with the pegboard all the pieces were safely back in the box. Then there was the time he accidentally upset the table. The old man was very apologetic about that, 'and wasn't it bad luck that you were in the lead?' Those chess games taught me that there is more to winning than being the best and that all is not lost until it's lost. Though I didn't enjoy it at the time, it is a lesson for which I have always been grateful.

Maitland Mackie had a proper dying at ninety. All six of his children gathered at North Ythsie. Everyone knew he was dying and there was no nonsense about 'When you are better Dad.' The old man, who had never been able to stand the sight of blood and had feared death all his life, was asked by his eldest daughter if he was afraid. He told her, 'I have no fear,' and she held his hand while he slipped slowly away, and a good while longer to make sure that he was quite gone.

Mary and Maitland were a remarkable couple, to whom justice cannot be done in this slim volume.

7
The Move to Buchan

Having read of my life with Maitland and Mary Mackie in their serene household in their prosperous corner of Formartine, you will appreciate that for one refugee the approach of peace in 1945 was not an unmitigated blessing. I was to leave my grandparents' house, and its kitchen and its garden and all my people therein.

I remember precisely when the terrible moment came. I was newly off to bed in Catherine's room. As usual the Old Folks had made nothing of the fact that my mother was coming home that evening from her war and that she would be reclaiming me. She put her head round the door, saw that I was awake and said, most warmly, 'Hallo Charles.'

The Refugee immediately began to cry. Uncontrollable deep sobs rent my small body. My mother rushed to me. 'Oh my darling. I know you're not crying because you're sad. I know you're glad to see me, really. When you're young and someone goes away you think they are dead and they'll never come back. When they do come back it is relief that makes you cry.'

That was good, sound child psychology, no doubt, but in this case it was rubbish. I remember my feelings perfectly. I knew the score exactly. I realised that this was the moment I had dreaded. I would have to leave the care of my granny, with whom everything was certain, and all my days were mapped out for me. Now I was going to rejoin my mother. I would be awarded choice in my life. I would have to decide whether I wanted to go to Aberdeen or not, what I wanted for my supper, whether I wanted to go to bed and whether I should eat up my porridge.

Old Northie had already settled his three sons in fine farms and in 1945 he made a start with the daughters, first by renting and then selling to my mother his wife's family home of Little Ardo, which he had bought from the laird in 1920. And the deal with my parents was done at the 1920 price. The 225 acres were bought and sold for £4,300. That included a range of farm buildings, which even then were no more than adequate, four cottar houses all with two bedrooms, running water and outside toilets. And a beautiful eighteenth-century farmhouse with outstanding views across Ythanvale to Formartine and Bennachie.

A good deal for my parents it undoubtedly was, but for the Refugee it represented a considerable drop in standard of living. The return to the regime of my somewhat bohemian parents meant that my routine was liberalised. I could no longer rely on there being any bacon rind to be tidied up after breakfast, and indeed I could not rely on there being any breakfast time or even any organised breakfast. The farmhouse, after twenty-five years of being an outfarm, was in a poor state and my parents either could not afford or didn't want the sort of luxurious furniture round which you could play cowboys and Indians when your octogenarian aunts came to visit. Then there was the very fact that Little Ardo was in Buchan.

There are those who regard Aberdeenshire as a gey bare place, and, if you have to generalise like that, I would have to agree. But I would qualify my agreement by pointing out that the Lowland part of what was Aberdeenshire before poor old Kincardine and Banff were annexed, is divided into three fairly distinct parts. Mar is the part which is almost Highland, and as such is beautiful and full of lairds and factors. But the real productive Aberdeenshire is the rest. The southern half is the Thanage of Formartine. It has many trappings of wealth. There are stately homes, Tower Houses, ancient trees, estates on which you can bag a pheasant or even, on the Lower Don or the Ythan, a salmon. Formartine is a relatively gentle land which has always given its people a fair return for their effort.

But cross the River Ythan and you're in Buchan. It is, in my father's phrase, 'the cold shoulder of Scotland which sticks out into the North Sea'.

It is flat and treeless with only the Mormond Hill, at 769 feet, to give a bit of shelter. There is nothing between Buchan and the North Pole, and when the north wind blows in winter you might as well be in the North Pole as in Buchan.

The good people of Buchan somehow got themselves on the wrong side in Scotland's Wars of Independence in the fourteenth century and that led to the Harrying of Buchan by that great Norman king, still revered in the rest of Scotland, Robert the Bruce. He and his men did their best, and by all accounts that was very good indeed, to leave not a house standing, nor a tree, and to kill every domestic animal as well as all the people of Buchan. The people and the animals are back but the trees never did recover.

There are bonny bits of course, like the parish of Old Deer, which was good enough to attract the first Christian monks in Scotland to set up an abbey there, but Little Ardo in 1945 wasn't one of them. It sat up on the top of a hill naked but for thirteen mature trees. Those were mostly plains, but there were two ash and three elms, and a hawthorn hedge which was quite blown to bits by the wind that came at it unimpeded from all directions but, thankfully, most often from the west. There was and there still is a much sturdier beech hedge around the farmhouse garden, but that was it.

Despite its rather gloomy topography, Buchan is nearly all arable. But that is not God's gift. There's no first-class land in Buchan; nothing that the fat cats of the Howes of Strathmore or the Mearns to the south would think fit for anything more than running a few sheep. That there's quite a lot of good second-class land is a monument to the toil of my forefathers. They were too busy toiling to bother with the aesthetic qualities or even the shelter afforded by trees. The mosses they drained and the stones they carted off, the seaweed and shells they carted on from the beaches and the lime from the quarries to sweeten the bogs would have beaten lesser men. But the men and women of Buchan stuck to their tasks and made a right job of nature's rather meagre bounty.

If the land of Buchan was infertile the same cannot be said of its people, who sent a stream of willing colonisers to the new lands of New Zealand and Australia as well as the Americas. Many went and prospered in Kent and Norfolk. And they sent to the Baltic and the Low Countries a stream of young men who could write and count to become merchants. There their training in work beyond reason, negotiation to the point at which weaker souls melted and their ability to count every ha'penny twice, saw the Buchan folk rise to the top like some distinctly earthy cream.

Yes, tough men and women made Buchan what it is today and I must tell you a little about one of them. His name was Geordie Gill and he was grieve at Little Ardo eighty-five years ago. In those days the wintering of cattle indoors, which was necessary not so much for the cattle as to protect the grass, was an even bigger problem than it is today. The mechanical digger and its little brother the muck loader hadn't been invented. That meant getting the tons of muck out to the fields started and finished with the graip. This could mean that a job which in 2008 takes one man a couple of hours might have taken two men working an eleven-hour day a whole fortnight of unremitting toil. Well now, Geordie Gill earned some kind of notoriety by loading ninety-eight cartloads of muck in one day, single handed.

Now if you agree that Geordie Gill was a hardy character, you'd be right. But to see just how driven were those who broke Buchan in from the moss, consider the reaction of his employer to the grieve's feat. My grandfather was absolutely dumfounded that the man, having done ninety-eight loads, couldn't be bothered to do another two for the century. But then Maitland Mackie was a past captain of the Auchnagatt cricket team and a century meant a lot to him. The fact was that, to Geordie Gill, there was nothing special about loading ninety-eight loads, perhaps ninety-eight tons. To the grieve a hundred loads would simply have been two more.

As well as being a legendary worker, Geordie Gill had a prodigious appetite for, and capacity to take, strong drink. It is a fairly common aptitude among the lightly blessed people of Buchan. It is said that Geordie once appeared after midnight at Grants, the general merchant's shop in Methlick, and hammered on the door. There was of course no one at the counter at that time of night but Pat Joss, the senior assistant, slept upstairs just above the door. He threw up his window, stuck his head out and enquired to know who it was and what the hell the errand was at that time of night. It was Geordie Gill. 'Oh, man, I've a stottie nae weel. I've tried aathing and it jist winna raise its heid. The only thing that could help would be a pint o whisky. Maister Mackie'll gie me the saik if I lose that stot without even tryin whisky.'

Not wholly convinced but not wanting to risk taking any share of the blame for the loss of a valuable steer, the shop assistant struggled up with a poor grace, and thrust a half bottle out of the door of the shop. Pat Joss was settling back into his still-cosy bed when he heard the unmistakable sound of a cork plopping out of a pint of whisky, followed by a faint gurgling and then the exaggerated sigh-cum-gasp of one who has made a good start to quenching a considerable thirst with whisky. 'Aye,' said the voice under the window, 'the stottie's feelin better already.'

Geordie Gill had long since left the farm when my mother, released from the war effort before Captain Allan, took me to Little Ardo to prepare a place for my father's return. We got the bus the six miles from the end of the North Ythsie road to Methlick. It was raining when we walked the mile and a half from the village where the bus terminated. We crossed the Ythan Bridge and into Buchan. Then along the Waterside, past Blackcraigs croft and the little ribbon of development which stretched almost to the Little Ardo road end. The farm road was, as it is still in 2008, steep and unmade. Half a mile up the Big Brae didn't seem too far or too steep. My mother was full of pride at going to take over her great-grandfather's farm. The small boy just trotted along. And there at the top of the hill, naked but for thirteen trees and standing stark against a grey November sky, stood our new home.

It was not an inviting prospect with its cracked and peeling harling and an appalling green painted and rotting wooden porch in the henhouse style. It quite obscured what turned out to be the handsome eighteenth-century frontage which was one day to earn the building C-listed status. What most impressed me about the house was the sound of it. The front door scraped open and each tread on the bare wooden floors echoed in a way that was quite foreign to the Refugee, who was used to the carpets of North Ythsie of Tarves.

My first impressions of outdoors in my new world were no more favourable. I remember Little Ardo in the early days as clearly as could be, and it is from the inside looking out. At the side and hard up against the house there is the nearer of two creosoted wooden sheds that were once for hens. The rain is sheeting off them, for there are no roans. It pours onto the unmade close, which is ankle deep in mud in the middle and to a five year old it is knee-deep in places where the staff have scraped the mud to the sides. Every few minutes a rat scuttles round from the midden to another eyesore, the sheddie whose front is nearly all glass and which has also been adapted for hens.

But for the boy who could mostly remember his time in Formartine, the most disappointing thing about the new home was the fact that there was not much of a hedge in which to look for birds' eggs. If the birds of North Ythsie were the blackbird and the green lintie, the birds of Little Ardo were the rook and the mournful peesieweep.

The house, though distinctly run down, was not altogether dissimilar to that at North Ythsie, having an eighteenth-century and a nineteenth-century half. At Little Ardo however, the two halves were both on the same level so, with no cellar, it was a poor thing of only three storeys. Again the back was occupied by a member of staff, in this case the grieve. In one important detail Little Ardo was the better planned; the Victorian part of the house, with its boring dormer windows, was definitely the addition and at the back. That gave Little Ardo the potential to be what it is today, a far more gracious dwelling than North Ythsie.

At that stage of my life such considerations didn't count. What did was that whereas at North Ythsie the back of the house was tenanted by the beautiful and sympathetic Miss Walker and her family, the back of the house at Little Ardo was occupied by the grieve of thirteen years and his family. They are a remarkable family, with whom I have the strongest bonds now and for whom I have the greatest respect, but that admiration got off to a slow start.

It may have been the sharing of the house, never easy even with your own main door and a locked door between the halves. But I think it was more likely that the Lows had managed fine for thirteen years and the Allans were an imposition. For whatever reason, two of the cattleman's sons and I fought like dogs with the grieve's two youngest boys and relationships with the parents were, at first, no more than polite . . . except the time I was alleged to have kicked Albert, who was my age and so ideal for fighting, between the legs. James Low came to the door seeking me to give me a hiding. I was glad then of my mother's protection.

Having been bowed in and out of the shops of Tarves with Mrs Mackie of North Ythsie, the reception in the village of Methlick was another come-down. We got on just fine in Hughie Buchan's sweetie shop, where the proprietor also ran a very small lending library. He said it didn't pay but people who came in for a read couldn't resist buying a sweetie while they were there. We got on fine at Duncan's Garage, and Grant's, the most central and the grandest of the three general merchants in the parish, and John Burr's the shoemaker. But we had no luck on our first visit to the Ythanview Hotel.

Tired out by her efforts to get the house ready for the return of the Hero, my mother decided to take her young son down to the village for a meal. It would be grand to have a meal cooked and washed up for her for a change. We walked down to the village, exchanging pleasantries with most of the people we met or who were tending their gardens which, at the tail end of the war, were full of fruit and vegetables rather than flowers. Digging for victory was more than a war-time slogan in Methlick, it was a way of life for most.

She rang the bell and we waited at the hotel door. After some time the door was opened by the proprietor Mr Macpherson, or 'Auld Mac' as he was known, a large figure who glowered at us from beneath heavy eyebrows.

'Yes?' he said without a trace of welcome.

'Good morning, Mr Macpherson,' said my mother. 'Charles and I have come for our dinner. The kitchen at Little Ardo is hardly ready and it will be so nice to have a change.'

'We're nae daein denners.'

'Oh dear. That's a pity,' said my mother. 'Well, perhaps we could have a sandwich?'

'Nuh.'

'Is there anything we could eat?'

'Nuh.'

'If I went along to Grant's and bought us pies could we have a cup of tea with that?'

'Nae teas.' At least he didn't finish by saying 'Have a nice day', or any other such artificial nonsense.

My mother's and my time at Little Ardo preparing for the Hero's return was exciting and my mother more than made up for her eccentric organisation in the attention she paid to me. In fact when I think back to how patient she was with the little boy who was never finished asking questions, I blush when I remember how ratty I was twenty years later when I had four toddlers of my own.

We did have a couple of bad moments. The first was when I found out about the sweetie ration. I knew that sweets and chocolate, as well as being the most delicious food imaginable, were 'bad for you'. As a good boy who was pretty enough to be a girl, I was always getting a sweetie from old ladies, so I accepted that my mother wouldn't buy me sweets. The sweets, like everything else, were rationed as the government tried to handle the inflation and shortages that surrounded the war. To get sweeties you not only needed money but everybody had a ration book and you had to tear or cut the relevant coupon out of that bookie. The ration was not great – roughly a bar of chocolate a week – so my mother, who had a very sweet tooth, as well as being very keen that her son eat nothing that was bad for him, had no difficulty in dealing with both rations.

Imagine, then, my wrath when I was told by the Low boys next door about my official entitlement to sweeties. I went to my mother in righteous indignation. 'Mummy, you've been eating my sweetie coupons!' She was suitably contrite and agreed that she was due me several years' sweeties. That was very fair of her and being able to share a sweetie brought us even closer together. However, with rationing in place, there were few opportunities for her to make good her debts, so she started making the most delicious tablet with whatever could be spared from our sugar ration. I got to claw out the pan and eat any crumbs that were left in the baking tray. So that crisis had a satisfactory outcome.

The second problem my mother had with her little darling was caused by her anxiety to obey her mother's rule that you should always tell the truth. The Truth Tells Twice rule, according to my mother, included Santa Claus. She thought telling children a lot of nonsense about an old man with a baggy red coat, a beard, reindeer and house in Greenland, undermined the bonds of trust there should be between parent and offspring. She was right. It led to quite a row, not because she lied to me about Santa but because she told me the truth.

My Uncle Mike Mackie and his wife Isobel used to put on the most wonderful Christmas parties at Westertown and there was always a Santa at those. Santa's entry was always exciting but I must just tell you about the best one.

At the stroke of midnight, which at these parties was about eight o'clock, Uncle Mike suddenly started shushing us, 'Do you hear that? Do you hear that?' he said. We did. From outside we could hear the tinkling of sleigh bells. The forty or so of us rushed outside where, on the front green, there was a searchlight. With it Uncle Mike swept the skies as though looking for German bombers. Then a voice that seemed to come from the roof shouted 'Ower here. Ower here.' It was a very Aberdeenshire Suntie with no trace of a Greenland accent. But Uncle Mike didn't seem to be as clever as we children and it took us ages, screaming advice to him, before he cottoned on and brought the light to bear on the roof of the farmhouse. And there he was, Suntie with sleigh and reindeer – just Rudolf, not the full team. They seemed to have alighted just beyond the top of the roof.

'Is this Westertown?' roared Suntie.

'Yes,' we roared back.

'Is that the lum?' said Suntie, pointing to a chimney at the other end of the house.

'Aye, that's it,' roared Uncle Mike.

With that the old gentleman on the roof said, 'Gee up,' to Rudolf and with much clanking of smiddy-made machinery, the reindeer and sleigh shoogled across the roof to the other end. There Suntie alighted and with bag on his shoulder climbed up and put one foot in the chimney, whereupon the searchlight went out and we all ran inside to greet him.

And the fun wasn't over. He hadn't appeared in the sitting room where the party was being held. Eventually, in an upstairs bedroom, we found him. Or at least we found his boot. Hanging down from the chimney and kicking was a wellington boot, which clearly had a foot in it and was the same make as my grandfather's, and a voice was saying, with more than a hint of panic, 'Help, I'm stuck.' After attempts to pull him out, Uncle Mike went for a mell hammer, which he just happened to have in the bedroom, and started battering the wall down. He managed quite easily to break into the chimney and Santa was rescued. He was quite dazed but was soon composed enough to come downstairs and give each of us a present.

I learned years later that my grandfather had been built into a false wall where he could dangle a foot down into the temporary fireplace. He had suffered terrible claustrophobia in there and I can't help thinking I have here an ideal plot for a short story. ' . . . the rich old man is persuaded to play the part of Santa and be built into the wall. But they use quick drying cement and leave the old gentleman to it. He is never discovered so they inherit all his wealth . . .' But what a nice man Uncle Mike was, going to all that bother to amuse a few kids.

My mother was coppit again. With evidence like that, how could she tell me that there was no such thing as Santa Claus and that it was just your grandpa or some other adult dressing up? 'You told me there wasn't a Santa Claus but there is because I've seen his boots,' I accused her. Despite my evidence she stuck to her guns.

Those little moral disagreements didn't spoil things for us and I think we both enjoyed those days together. Extra special were bedtimes, when she read me a story. And Jean Allan didn't read to her son about Noddy and Big Ears. She read me the classics. I had two favourites. First was Charles Kingsley's translation of *Homer's Heroes*, which I confused as 'Homesby's Heroes'. At the age of six I knew all about Jason's voyage with the Argonauts to capture the Golden Fleece, Ulysses' narrow escape between Scylla and Charybdis (though I can't remember which was the six-headed monster and which the whirlpool) and Perseus killing the beautiful Medusa, who had a wonderful head of snakes instead of hair. And as if that wasn't culture enough my next book was John Bunyan's *Pilgrim's Progress* following Christian's progress toward heaven through the Valley of the Shadow of Death, outstripping both Obstinate and Pliable. Perhaps it's a wonder I slept at all after a few pages of those, but I really did love them.

8
The Yulls

I fear that I should start recounting the story of my maternal grandmother's people with the rather unsavoury business of why, when all the other people in the world whose name sounds like Yool spell it either Yule or Yuille, the Yulls should spell their name differently. Well the truth, as I have always been told it, is that my granny's people were Yules until a terrible row led her great-grandfather to tell his siblings something to the effect, 'If that's the way o't, I'm for nae mair to dae wi you. In fact, I'm nae even gaun tae spell my name the same as you', and started signing 'Alexander Yull'.

Alexander's son William did better. In fact he covered himself with glory and received the Royal Highland and Agriculture's gold medal for what he called his 'experiments' while farming the Mains of Fedderate in the parish of New Deer, some seven miles farther into Buchan than Little Ardo. That was a Herculean effort in bringing fifty-one acres in from the moss and improving the remaining fifty acres of his farm. And while much of this volume is hearsay, the details of William's experiments are all beautifully documented. I have a copy of the submission to the Royal Highland and Agricultural Society which won him the gold medal. I can give you no more than a flavour here, but it was some undertaking.

William started with 2,047 yards of open ditches four and a half feet deep, fourteen inches wide at the bottom and seven feet wide at the tops, to define his fields and run the water off, and into which he could direct his drains. He then dug 5,300 yards of drains. Those were four and a half feet and filled to a depth of three feet with stones. The surface was covered by 30,000 horse-cartloads of clay and gravel, each one hand-loaded, which was mixed with the moss and what came out of the drains, to make his topsoil. It would have been a huge endeavour with diggers and 'dozers, but this was all done by hand with spades and wheelbarrows. I can only imagine what it was like to cast a drain seven feet wide by four and a half feet deep, but the thing I can most nearly identify with is the hand-digging of land which should have been ploughed. Well, William Yull somehow got one field of ten acres dug and levelled by spade.

And the old man deemed his effort worth it. He trebled the yield of oats from the first to the third year of his improvements . . . and remember that what he started with was deep moss, much of which had been cut for peat and stone quarries, leaving deep water holes. To get rid of the water he had to fill the quarry-holes with stones until the water level reached that of his drains. All those loads of stone with which he filled the drains had to be carried by hand or wheelbarrow because the moss was too soft to support the horses.

Now, there is a piece of family folklore about William Yull and his improvements and what happened next. The story goes that the Laird of Brucklay, who was William's landlord, rewarded his tenant by offering him a new lease. Fine but, as the farm was so much improved, he would need to pay a considerably increased rent.

Having paid for the improvements once, spending no less than £352 11s 10½d of his own money, William Yull was damned if he would pay for them again. Luckily, the story continues, Lord Aberdeen, along with the rest of the county, had heard of the Fedderate improvements and the gold medal. He saw the chance to get a really good improving tenant and offered William the vastly better, and almost twice as big, farm of Little Ardo in the neighbouring parish of Methlick. That is how in 1837 my family came to Little Ardo.

About the year 2000 I got the chance to put that story to the then Laird of Brucklay, Andrew Dingwall-Fordyce. He was intrigued and promised to find out what he could about what his forebears were alleged to have done. It didn't take him long. In fact, at the time of William Yull's improvements the Mains of Fedderate was already part of the Aberdeen estate, not yet having been acquired by the Dingwall-Fordyces. The next few generations of my family will now be told that as a reward for his efforts in New Deer, William Yull was offered a better farm by his laird. It isn't as good a story but I fear it is nearer the truth. And at any rate, sitting as it does on the south-facing slopes of Ythanvale, with its free drainage and spring water everywhere, he got a farm that 'would neither droot nor droon'.

In contrast to the rather boring rectitude and remorseless success of my Mackie progenitors, the Yulls were a much more chequered bunch who survived much closer to the truths of the bare land from which they had sprung. For example, if there was ever a poacher among the Mackies none was ever known for it. But John Yull was. My maternal great-grandfather, William's son and successor as farmer of Little Ardo, was a delightful man. Of more than average stature, he had a bowler hat, wore heavy tweeds, a solid gold watch chain and whiskers. But much more than that, he had a cavalier spirit and a great sense of fun. One of his most famous exploits was when, for a wager of £5, he rode his gig home from Ellon cattle market with neither reins nor traces. He held on only to the tail of his great mare, Clatterin Jean. Jean's tail must have been stuck on better than Tam O'Shanter's grey mare Meg's, for the good horse pulled gig and master home and the bet was paid.

I suppose it was for more fun that the old man poached salmon. It certainly wasn't for profit, like those who come in the night to empty the Ythan nowadays. John Yull just took one for the pot, as many did in those days when times were hard and diets meagre. So unremarkable was the crime that there used to be a rule imposed by the single men on their employers that they not be given salmon more than three times a week. But John Yull wouldn't have taken salmon out of necessity. He farmed one of the bigger and better farms in the district and, though he was no more than fair as a farmer, he made a good enough living out of his gift of the gab, for my great-grandfather was an auctioneer, and a good one.

When William Duthie, the shorthorn pioneer, held his first on-farm sale of bulls at Collynie, he naturally gave the job of selling to John Yull, and not just because he was a neighbour. John Yull was in great demand throughout the county. He was able to keep the sales moving at a speed which became legendary, with the aid of a stream of banter. I offer one example and leave with you the question of whether it is fit to print. John Yull was asking the exorbitant price of half a crown for a chamber pot when a lady, standing at the front where she could see every detail of each item, shouted out, 'There's a hole in't.' 'There's a hole in you tae wifie, but you're nane the waur o that,' replied the auctioneer as quick as a flash. 'Now, who'll gie me twa shillins for the chantie?' Some said they never liked to let their attention stray when John Yull was in the box. You never knew what he might say.

One bright, crisp morning in May about the year 1892 John Yull was after a fish in the Mullart's Pot just below the farm of Mill of Ardo. Though the river was quite high the water was clear and he could see deep into the pool, which was just on the bend in the river where the current had, over the years, hewn a smooth swirl out of the rock. Settled there, my grandfather saw the fish he wanted. It was a grilse weighing about six pounds, the sort of fish a rod-and-line man in the pub would call 'a ten-pounder'.

Now most poachers in the Ythan, and its tributary the Little Water, worked at night with a gas lamp and a gaff or even a graip from the byre. But John Yull considered that a cowardly way of doing and he didn't like the way the gaff spoiled the look of the fish. He used the light of the sun and a snicker. That's a wire snare more often used to catch rabbits, hares and the occasional cat, with a long piece of strong string. He fixed the snare to the end of a six foot long stick and tied the string round his wrist. Then he pushed the stick into the water and looped the snare over the fish's tail. That was the tricky bit, for one touch and the fish would be off. But there were no such problems that day. The snare over the tail of the grilse, my great-grandfather let go of the stick and yanked the string (upstream lest he pull it off the tail) and there thrashing on the bank was a beautiful fish, fresh-run and silvery.

Now, as luck would have it, the water bailiff came along at just the wrong time. He couldn't see exactly what was happening because of the way the bank of the river sloped down to the water, but he didn't have to. He recognised John Yull and knew of his reputation. Too late, John Yull saw the danger. What should he do? It was both too late and beneath his dignity to run and the fish was now dead and could not be thrown back for it would clearly be seen floating belly-up in the slow-flowing river. Luckily, my great-grandfather was not alone on the riverbank. His daughter Mary, aged thirteen, was there too. Beautiful with her long auburn hair held back by a white ribbon, she sat on her rather overweight Highland pony, Donald. It was then that Mary Yull showed for the first time that composure which was to become legend in the North-east over the next seventy years. She bade her father hand her the fish and, after concealing it in the flounces of her long black dress turned Donald and started up the hill to her home, bidding the bailiff a civil if slightly condescending, 'Goodday Baillie'. The bailiff was no fool. He knew fine where the fish was. But those were gentler days and there was no way he was going to question what was among Miss Yull of Little Ardo's petticoats. I have some sympathy with him for, though I know it happened, I just can't imagine the all-wise, all-powerful, all-virtuous paragon who looked after me during the war going poaching and making off with the kill under her skirt.

It wasn't that narrow squeak that stopped John Yull's poaching. It was a piece of brilliant man-management by the laird. He was determined to stop his illustrious tenant's activities, though I really don't see why. Salmon were so plentiful in those days that they used to joke that when the water was low you could walk across the river on their backs without getting your feet wet. But lairds do have their pride of ownership and the poacher must be stopped. The poacher was invited to a day's rod-and-line fishing with the laird's fancy friends. There is no report as to whether the poacher caught anything but John Yull had been a great success socially and, having seen the other side of the fence, he never went poaching again.

The end of his career as a poacher was not by any means the end of fun for John Yull. He was by all accounts wonderful company and was best man to no fewer than thirty-six of his pals. In common with most of the agricultural industry in his time, he was a social drinker. Just how sociable he was is indicated by the records of John Grants, the general merchants of Methlick. Those show that on Hogmanay of 1889 John Yull laid in a stock of two gallons of 'aqua' against the possibility that friends might happen in bye to wish him a happy New Year. The same records show that on 1 January 1890 the farmer of Little Ardo was down to the village for another gallon of aqua.

And it wasn't just at the turn of the year that the social mores of the time forced my great-grandfather to strong drink. For, as well as social drinking, there was that splendid institution, the business dram. If you sold a fat beast at Ellon Market, a dram with the auctioneer was a fine way to seal the deal, to make sure that he would remember you should there be any question arising from the sale . . . should the beast turn lame or, if it were a breeding cow, should she turn out not in calf or, if a heifer, turn out to have had a night out with the boys across the burn. Buyers as well as sellers might offer a business dram, so an auctioneer who sold fifty lots in a day was on a bit of an assault course.

Luckily there was no breathalyser when John Yull was an auctioneer and in respect of getting home from the mart he was doubly blessed. It would be late when he climbed, sometimes unaided, into his gig and took the road to Little Ardo. He would make the loud kissing noise with his lips that told a good horse it was time to go, and then settle down to sleep. As long as the mart was in Aberdeenshire the excellent Clatterin Jean would get him home. Many were the times my great-grandmother heard the horse nosing at the stable door. She would rise, waken the old man and bring him in. She would then stable and feed the old mare. What a horse and what a wife!

Another side of John Yull's character, his unflappability, is illustrated by another of our family's favourite stories. John Yull's wife, my maternal great-grandmother Jane, liked nothing better than a nice juicy orange at the fireside. She would peel the fruit into her skirts and when she was finished, gathered her skirts and flicked the peel all at once into the fire, where she enjoyed the flare-up that caused. Sadly, on this famous occasion, the false teeth which she had taken out to facilitate the full enjoyment of the orange were in her lap among the peel. 'Oh John ma teeth!' she shouted out. But John could see that the teeth were gone. He was neither going to risk his hand in a lost cause nor was he going to cry over spilled milk. He just moved the pipe to the side of his mouth and said with neither sympathy nor anger, 'That's a fine fat stot away up the lum.'

John Yull believed in having the best of everything, whether the tenets of sound finance dictated that he could afford it or not. If anything new came out, Little Ardo would have it. It is one of our proudest boasts that when the entire rest of Aberdeenshire, including Lord Aberdeen himself, was still 'gaun tae the furth' (doing their business outside) or messing about with pails and chamberpots, Little Ardo had a water closet . . . a wonder in its day.

When Mary Yull had left the southern tip of Buchan and moved to the far grander farm of North Ythsie in the doucer environment of Formartine, she was quick to take the part of Little Ardo and not let the Mackies get too grand with her. On one occasion when she was reminding her husband that at Little Ardo she was used to having a flushing water closet, he is said to have replied, with heavy irony, 'Aye there's aathing there but siller.'

Then there was Isie, John Yull's illegitimate daughter, who was brought up by her mother and financially supported by her father. When the day came that Isie's mother got married, her prospective husband didn't want to raise somebody else's bairn. And as he already had a house, he had no use for the furniture Isie's poor mother had in her cottar house. And so it was that my grandmother, at the age of seven, acquired a sister three years older. Suddenly one Sunday, there she was in the close, sitting on top of a pile of furniture on a single horse and cart.

Mary, my grandmother, was not pleased. Isie was a daily reminder that her father had not lived up to the standards which his legitimate daughter espoused. There is no word of how old Mrs Yull took the arrival, though there is a suggestion that she was not forewarned. At any rate someone must have been nice to Isie, for she turned out well. She got a job in domestic service, saved her money and invested it under the direction of a grateful employer. My mother and I visited her in a very comfortable private nursing home in Nairn where she told my mother that there would be enough money to see her out. 'And if there's ony left you can hae it to sort the brae at Little Ardo.'

Indeed there was £2,000 left, a considerable sum in the mid fifties, when Isie died and my parents did get it. It is a source of some shame that, though the shortest road to Methlick was metalled down as far as the 'single hoose', it has been thought better to bypass the Brae, which was very steep, and take the longer way round by the Loans.

John Yull died in 1905 with three of his children – those who had not yet left for the Colonies – about him: Mary, Isie and George, known as Dod, who followed his father in the farm but was eager to be off to Australia. Family folklore only includes one image of John Yull's dying. The old man kept his two daughters busy for weeks going for one last drink. But the man who had needed three gallons of whisky to see in the New Year of 1890 did not send the girls to the whisky pig. He wanted just one last drink of the wonderful cool clear water that seemed to come inexhaustibly from the well in the close. 'Jist a last drink fae the wall,' he would reply when they asked him if there was anything they could do. Unfortunately for the girls, his appreciation of the water was much better than his judgement of how long he would live. It went on for weeks until they were heartily sick of trekking outside for 'Ae last drink fae the wall'.

Eventually the legitimate one said she had had enough. She would get it from the tap in the back kitchen. But with Dickensian irony the illegitimate daughter insisted that, after all the old man had done for them, the least they could do was to get him a last drink – no matter how uncertain its timing – from the well of his choice. After John Yull's death the laird (or his factor) was determined to offer the tenancy of this fine farm to another. No doubt he could have got a bigger rent, but Mary, my grandmother, took the matter in hand. She cycled the twenty-six miles to Aberdeen to see the laird's legal people and somehow persuaded them that they could not do that and that her brother George should have the tenancy. That was done and old Mrs Yull was able to stay on for a few more years.

They were not happy years for the widow. Her daughter Mary left for North Ythsie and marriage to Maitland Mackie in 1906. John was already in Australia and Jimmy in America. Isie, who had been no relation of hers anyway, had left to go into service. George was desperate to be off to the New World too. In 1911 he gave up the tenancy, rouped out, selling all the stock, machinery and household effects, arranged for his mother to move in with Mary at North Ythsie, and Maitland Mackie took over the tenancy.

I have this pathetic vision, which had its equivalents all over Scotland in those days when we peopled the Empire, of old Mrs Yull standing at the top of the Big Brae that leads down to Methlick and watching as George, the last of her sons left. The last thing she said which anyone remembered was 'If I could just hear fae Jimmy afore I die.' The others all wrote, but Jimmy had just disappeared down the brae.

That was the end of the Yull name at Little Ardo but, through the female line, we are still here. In 2008 there isn't a single Yull in the North-east of Scotland phone book, but there are seven in the book for Perth, Australia.

9
The Lie of the Land

The farm lies at the bottom of a slope of some three miles which takes you from the top of Bennagowk down to the River Ythan that is the boundary which separates the Thanage of Formartine from Buchan. From the top of the Hill of the Cuckoo (as Bennagowk translates to Queen's English), at 579 feet, it falls to 90 feet above sea level at the river. It still has almost ten miles to go to the start of the tides at Ellon, so the Ythan makes sedate progress in dividing the two best bits of agricultural Aberdeenshire.

The farm of Little Ardo was laid out over 200 years ago. Then it comprised one field either side of a straight road running due north from the river and extended to 150 acres. I said a straight road, but that isn't quite right. The dykes which define the area allocated to the road are straight but they are an average of forty yards apart. It is rather like an old drove road, being wide enough for a large herd of cattle or flock of sheep to graze its way along and even meet another such herd without becoming impossibly entangled. The road itself is properly made of burn stones washed, and often washed out, by the water which cascades down it when it rains. And it wends its way upwards between the dykes, winding wherever necessary to take the worst of the steepness off the hill, which runs to a gradient of one in four.

These 150 acres have been farmed for a very long time. Three thousand years ago the Beaker people were around and we would be sure they had been at Little Ardo if we had just found one relic. It wouldn't need to be a stone circle or a burial cyst; even an arrowhead would be proof enough for us. And yet there is nothing. Still, considering those farmers' eyes for fertility, and the consistency of the soil, which had clearly been turned a thousand times, we are convinced the Beaker people were the first of us on the little hill. There is further evidence for that, because such relics have been found a mile upstream on the other side of the Ythan on the Braes of Gight, and there is a Pictish settlement on the other side of the Methlick to New Deer road, not half a mile from Little Ardo's land.

Those 150 acres have been added to over the years. There was a time when the tradespeople of the village needed a field for their house cow and, much more important in the days before the internal combustion engine, to graze the horse that pulled them around the countryside in their gigs, and to make some hay to see them through the winter. Before the First World War there were over thirty such horses on the village lands. The coming of the motor car finished all that and many of those fields were added to Little Ardo. The Free Kirk Glebe was a grand affair. When the Kirk disrupted over the issue of how much power the lairds should have over matters divine, Lord Aberdeen showed that he wasn't annoyed by giving the Free Kirk not just a site for their church but twelve acres of a glebe. In the middle of the nineteenth century that was sufficient for the full-time employment of a man. So when the Methlick congregation of the United Free Church decided to rejoin the Kirk in 1933, Little Ardo acquired four little fields totalling twelve acres. Then came the doctor's field, which was some six acres, and the Waterside, a four-acre field down by the river. With that came the title to salmon-fishing rights which could have saved my poaching great-grandfather a lot of bother had they been added sooner to Little Ardo.

The family farm grew by absorbing little crofts which mechanisation and rising standards of living made uneconomic as the twentieth century progressed. There was until the 1930s a blacksmith's to the north of Little Ardo and he had fifteen acres in two fields, one which got all the muck and was very fertile and one which was very wet. When the blacksmith gave up the laird added that to Little Ardo. And when I was a boy there were ruins of another croft to the east side of the farm next to the Glebe and when that had been given up the laird again added it to the little farm on the hill. Seven acres of very rough common grazing were added in 1923 and old

Maitland Mackie had planted trees in that just before the Second World War.

We also have one of the village lands, called the Admiral. Sadly the Admiral was not one of the village trades, but he was the tenant who added grace to his retirement by building a quaint pink bungalow with a daft little turret on it, on a rocky bit overlooking the village. Whether he grazed a horse on it or not I don't know, but the field next to his house still bears not his name but his job title. There are stories about the Admiral which still keep him alive for us. He used to get a load of muck for his garden from Little Ardo each year and the grieve, Geordie Gill, would take it down in person. That may have been because the old man was good company but it was also because he always gave the horseman a good dram. It is said that they had an annual ritual in which the Admiral gave the grieve a big toddy glass, which would have held three quarters of a pint comfortably, and started pouring saying, 'Say when'. It is said that Geordie waited until the whisky was lipping the brim before saying, 'Now Admiral.'

So that was the farm to which the Hero returned in 1945. He estimated its extent at 225 acres, which didn't include the roads and steadings. John Allan added no land, so the medium farm of 225 acres in 1945 was a small farm indeed by the time I took over in 1973. I was able to add the joiner's croft when the last of the Patersons, the firm of millwrights who had had it since deep into the nineteenth century, gave up and left for village life in Tarves, seven miles away.

Having the north slope of the valley, Little Ardo has the first of the sun each day, which has always meant it was an early place at sowing or harvest. Being on so long a slope it had ample spring water, so it would never drought, and being on the slope it would never drown either. It was a good farm by the standards of the area and of a size that in 1945 could have been called economic. The story of the farm from then onwards is of how, as the size necessary for economic farming grew, and as the amenities appropriate to the staff and the farmer grew more and more expensive, the farm which was once big gradually became small and the 'gentleman farmer' became, at the last, 'the crofter'.

10
The Allans and the Robertsons

I have dealt at length with the forebears of Maitland Mackie, my grandfather, who had farmed Little Ardo most recently. And the Yulls, my granny's people, who were the first of us to have farmed Little Ardo. But now another farmer was to come striding up the brae. What about his people? What about my other grandparents? What about the Allans and the Robertsons?

'Every time a child is born a ploughman leaves the village.' That rather cruel saying was an obvious exaggeration, but it is what happened in John Allan's case.

His father's name was John Robertson and his mother's name was Fanny Allan. Both were from Aberdeenshire farming stock and both were fee-ed on farms. John had been disappointed by Fanny's friend, who had gone off to the Empire and left him. On the rebound from that he took up with Fanny. When she became pregnant he left for California, where he is reputed to have married and had seven daughters but no other son than the one in Scotland, with whom he had no contact whatsoever and may not even have known existed.

My paternal grandmother, who was in service away from home, somehow managed to conceal her pregnancy from her parents. She managed to have the child and have it taken into care in 'an institution' before emigrating to Canada. There she married Fred Willows and had three other children. Meanwhile back in Scotland, Fanny's parents had heard about the baby. They found him in the institution. They took him home and gave him their name and a wonderful childhood. I don't want to spend too much time describing the Allans for two reasons: firstly, they played little part in the story of Little Ardo; secondly, the Allans and Bodachra are so well written up by a far more accomplished wordsmith and knowledgeable historian than myself, in *Farmer's Boy* by my father, John R. Allan, published in 1934 and reprinted every few years, with the tenth edition scheduled for 2009.

There is one item involving the Robertsons in the folk memories of my family, however. When he was old enough to remember it, John Allan was taken to see (or to be seen by) his other grandparents. He remembered his Granny Allan saying, 'We called him Johnny Robertson – it seemed only fair.' I remember thinking when I was told that, that what old Mrs Allan had meant was that it was fair to the Robertsons that they get a mention. But I wonder now if she was really saying it was only fair that the boy, who might have been christened a Robertson, should at least have his father's name somewhere on his birth certificate. He was given half a crown and that was it, as far as the Robertsons were concerned, except that when the boy became a broadcaster and a writer he carried the name forward with distinction, and John now has a grandson, John Robertson Yull Allan, who is carrying that weak association forward again.

So John Robertson was brought up an Allan on their family farm of Bodachra beside Scotstoun Moor just outside Aberdeen. He could have done an awful lot worse, for the Allans were considerable people.

The Allans can be traced back to one James, who was hanged for sheep-stealing. They seem to have been a roistering and gregarious crowd who might have got on well with the Yulls. Though there is no record of any meetings, there must have been some at the cattle markets and horse fairs where the farmers of the North-east met to work and to play with a ferocity that is uncommon.

The great Allan ancestor was Charles Allan, and I like to think of him meeting John Yull of Little Ardo who drove his gig home from Ellon market holding on only to his horse's tail. Would John have taken on Charles Allan's famous challenge? He used to offer to race the fastest runner in any boisterous company, carrying the heaviest man on his back. It was a handicap race, of course, and Charles's skill appears to have been to so organise the handicaps that everyone was convinced that he couldn't possibly win, so encouraging people to bet against him – and then astonish them and clean up. It is just as well that no one ever tried to make the same offer when Charles Allan was available for the post of heaviest, for he weighed thirty-two stone. It's not the ideal bulk for a sprinter, even without the heaviest man on his back, but our family folklore hasn't just told that story twice. It has been told a hundred times and each time it has come out more or less the same.

Another Allan family story also refers to Charles's extreme bulk. He once found two teenage nephews at some sort of devilment about the farm, caught them (both) and carried them to the washing line where a pair of his enormous trousers were drying. He then lowered one loon down into each leg of the trousers and then beat them soundly with his wife's carpet beater. This is an excellent story, but it does try our definition of the truth. How did a thirty-two-stone man catch not one but two teenagers simultaneously, and what sort of clothesline could support and hold two youths who were anxious to escape a beating? But that's how the story is always told, and the truth tells twice.

According to John Allan, my father, Charles was the only one of the Allans who had 'anything of greatness in him'. It was not only in his love of Bodachra that he was great. A tenant farmer who had to toil for a living, he was one of the best-known figures in the City of Aberdeen – he took his Friday dinner at the Red Lion at the same table as the Lord Provost and the Dean of Guild and carved his beef off the same joint as men who could have bought and sold him a hundred times and was treated as their equal – at least. I have seen a volume about Aberdeen in the nineteenth century which described a great procession in the town. There were pipe bands, a company of the Gordon Highlanders, decorated floats representing the incorporated trades, the Burgesses of Guild, The Master Butchers, Bakers and Candlestickmakers, and with no other identification than his name, Charles Allan. He had been an attractive character, as John Allan made clear in *Farmer's Boy*:

> A roistering young man who seemed destined for the devil. He liked good living, dancing, drinking, blocking chimneys, courting every pretty girl within miles. He was extravagant, splendidly generous, an indifferent businessman and therefore poor. But his religion was his love of Bodachra and he

> obeyed the old peasant's law; he never let his farm know when he was hard up. No matter how hard the times were he never let the land go short of guano, bonemeal or lime. Charles it was who bought the family plot and paid for the gravestone that stands to this day in the graveyard of St Machar's Cathedral. That may sound very grand to you and it does have great resonance for those of us brought up in the North-east, but the cathedral was Bodachra's parish church. Anyway, Charles Allan left a better memorial which stands to this day. He brought in sixty acres from Scotstoun Moor to add to the sixty acres his father had added to Bodachra before him.

The great ancestor among the Allans had seven children with his wife, Susan Baird. He outlived his wife by eighteen years and died in 1874 aged only sixty.

But I haven't said anything about the great Allan ancestor's father. Well he was also Charles and he was the first Allan to farm at Bodachra, yoke his gig for Kittybrewster market on a Friday and attend St Machar's Cathedral on a Sunday. At that time Bodachra was no more than a windy pasture between the moss and the moor. At first he had had some shelter there and some peats to cut for his fire, but the rent had bought the first Charles Allan only some very sparse grazing for no more than a beast or two and some hens. But Charles cut the first sod with the spade, made great dykes to define fields and consume the stones, dug deep ditches and laid subtle drains that would suck the acid water from the moss. Then he put in the plough. He went to the beaches at Balmedie with his horse and cart for seaweed and for shells to add the missing minerals. He broke the moor and the moss into a boulder-free loam that would sustain cash-crops though they brought him precious little cash. When he died Charles was still only a tenant but that didn't matter. The land was what did matter and he had broken it, tended it, loved it and reaped an increasing harvest from it. When the first Charles Allan among us at Bodachra died in 1836, he had performed the peasant's first duty – he had left the land to his son in better heart than he had found it. In fact he left a farm where there had been nothing but blasted heath. The record shows only three sons, but considering that there were eight years between his marriage in 1806 and the birth of the great ancestor Charles Allan in 1814, I am sure there were several older daughters as well. But of course women hardly counted in those days.

As we have seen, his son added another sixty acres from the moor and the moss and that was the farm to whose tenancy Alexander Allan succeeded when his father died in 1874. Alexander took on responsibility for the farm and he also took responsibility for a wife. He married Fanny Sim, the daughter of a neighbouring farmer, when she was twenty and he was thirty-four. They had eight children and that was the family that rescued my father from 'the institution'.

Of the Allans' eight children, one had died at birth and the first, a boy called Andrew, had died in 1896, aged eighteen. That left three boys and three girls. In 1906, when the old people brought young Fanny's baby home from 'the institution', John Allan was only nine years younger than his Uncle Alexander, so he wasn't perhaps too far out of time for a big family in those days. He loved his grandparents and they seem to have loved him back. They certainly gave him the framework for happiness.

The house and garden were the province of his grandmother. She explained things without worrying him, and so long as he didn't break anything, eat what was forbidden, or make an intolerable mess, he had the run of them. The front room at Bodachra was to the boy just like the heaven she described for him but on a smaller scale. And if heaven was in the house, the Garden of Eden was outside the front door. There old Fanny Allan presided over a riot of fertility all summer and, in season, John Allan found a dish with seven strawberries on it every morning on the window sill of his bedroom. The old woman shared John's room for his first six years at Bodachra and when he was seven he moved in with the old man for care and education in more manly things. The old man liked to have the boy with him as he toured the farm on his walks or the countryside on his gig. And it wasn't just how the land was won and how it could be kept that Old Bodachra passed on to his grandson, he told him all the gossip of the countryside and gave him the taste for a good story which he never lost, and from which he was to make a good living much later.

Something of the old couple's way of bringing up their daughter's son can be gleaned from a couple of stories my father used to tell me when I was the same age he was when his grandfather taught him the way of things at Bodachra.

First was the sad case of his experiment with his granny's eggs. Old Mrs Allan had a way of telling whether hens' eggs were being incubated below a clocken hen or were found laid away were gorbled or not (whether they had been fertilised). She put the egg to her ear and then pronounced them fertile, or for John to break against the steading wall. If they are close to hatching anyone can hear the chick starting to break its way out, or trying out the limbs it will need when it gets out. But the old woman had an uncanny ability to tell whether they were fertilised at a very early stage. Fired by the spirit of enquiry, my father thought to try an experiment. Could he tell whether eggs were gorbled?

Sadly, the young scientist had not seen the difficulty that it was no good just guessing which were fertile and which were not. You had to verify the results. The only way to do that was to break them. The old woman had a big poultry enterprise and she had a whole henhouse full of mother hens sitting patiently on about a dozen eggs apiece. When his grandmother came in and found him as he was breaking the last nestful, he looked up at her with his most innocent smile and said, 'I wis tryin tae see if there wis ony chuckies in them.' She looked at her disconsolate hens and the gigantic omelet on her henhouse floor and said sadly, 'Ah weel. We fairly ken noo.' Other than a look that would have challenged hell to freeze over, there was no retribution.

The young scientist didn't fare so well the time he accidentally sat on his grandfather's very smart, four-cornered bowler hat. It was reserved for Sundays and for funerals. It had been a great expense and had been bought to see the old man's time out. He had taken it off as usual and laid it on the pew beside him. The boy had accidentally squashed it. Now that would have been bad enough had my father owned up, but he chose not to. Already in a hole, he determined to keep digging. Desperate to shift the blame he even tried to shift it onto an 'ancient disillusioned spinster' who sat quite near but not nearly near enough to have done any damage to the hat. His grandmother advised him that his punishment would be less if he went and owned up but John Allan had told his lie and decided, unwisely, to stick to it.

On the way home from church the lie was still being held. The young liar usually sat between his grandparents proudly at the front of the gig but this time the back seat was let down and he was sat on that with his grandfather's grimmest look. By the time he got home in miserable silence John Allan was ready for repentance but no one mentioned the hat. Not even at dinner was it brought up and he couldn't eat a thing. By the time they rose from the table John could stand it no longer. He found his grandmother, buried himself in her skirts and proclaimed his guilt – and was forgiven immediately. But that was the easy bit. He would need to go and apologise to his grandfather and his granny knocked back his cowardly suggestion that she convey his apologies.

'But he'll maybe dae something awfu tae me,' said the young man.

'I wadna be surprised at that,' said the old woman. 'You'd better awa and get it ower.'

So he went trembling into the parlour where the Old Man was having a post-prandial doze by the fire, the ruined hat on the table beside him.

'Weel, fut is't?' he asked grimly.

'The hat – it was me – and I'm sorry.'

'Ye devil. Come here.'

He went.

'Bend doon.'

He bent.

'Noo tak that,' said the Old Man and caught his grandson such a hoist on the backside with his Sunday boots that he landed in a heap at the other side of the room, shaken but not really hurt.

'Noo than laddie, awa and get my staff and we'll gang doon and hae a look at the stots on the moss.'

Honour was satisfied and that was the end of that.

Young John, or Jakie as he was nicknamed, fared less well with his Uncle Charles, with whom he already had fragile relations. The problem this time was Uncle Charles' piglets and another scientific question. He knew that if cats were thrown in the air they would always land on their feet. But what about piglets? Charles had a sow with a brand new litter of eight piglets; pink, pretty and accessible to the young scientist. Horses and cattle were far too big to be thrown up into the air to see if they would land on their feet, but piglets weighed no more than three pounds. The scientist knew that cats had to be thrown a reasonable height in order to do the necessary twisting to allow them to land feet first, so likely a piglet would need the same. So he threw the piggy as far as he could in the air. And the experiment was a success. It was clear, as the piglet crashed squealing on its back, that pigs cannot twist like cats.

Sadly for the scientist, Uncle Charles happened into the piggery just as the piglet was spiralling skywards. He was not impressed by the experiment and was uninterested in discussing it. He grabbed his nephew and here's how John Allan described what happened next:

> He let out a bellow of rage and fell on me hip and thigh, hauled down my slacks with a ruthless hand, so that the buttons flew in all directions, laid me across his knee and let me have it. When he had avenged his little pig, Uncle Charles gave me a final grand smack that echoes in my head to this day, set me down and left me there. And there I stood, my world about my ears, my little slacks about my feet, bawling out my anger and humiliation to the unheeding afternoon while the little pigs rooted and squealed in the pen nearby. I don't know if my mind or my buttocks hurt more. After a little I began to move towards the house greatly impeded by my sobs and the trousers hobbled round my ankles. Every now and then I stopped for a grander roar, then moved on when I had exhausted it. I had got as far as the middle of the close when the anger, humiliation and pain overcame me. I could go no farther, but stood there with my knuckles crammed into my eyes, my trousers dragging behind me, and my pink shirt fluttering round my tender buttocks.

That was where his protector and Aunt Susie found him when she returned from the fields. She asked no questions and offered no sympathy, but pulled up his trousers, tucked in his shirt and said, 'Noo than, come and help me bake bannocks.' Then Susie continued with the Bodachra education by explaining to John Allan that his spirit of enquiry had to be tempered with common sense. It was a pretty humane approach.

John Allan's upbringing wasn't all easy. There was the question of his birth. My father was hardly at fault for the circumstances of his birth, but it was a strange fact about those times that the children were made to suffer for the sins of their fathers and mothers. Some of it was institutional. He had to take his birth certificate to school and it had 'illegitimate' scrawled across it and no sign of a father's name. Somehow the boy felt that. The teacher looked at the birth certificate and oozed disapproval at the little round boy. And he felt the way strangers visiting looked at him as though to say, 'Ah yes, I know something about you. You'll be Fanny's loon.' It was a trouble to him at Bodachra and it followed John Allan all his days.

Then there was his Uncle Charles. He was very hard on his nephew, who might have been like a little brother but certainly wasn't treated like one. That unkindness wasn't so much physical (the incident with the piglets certainly doesn't count – that sort of thing soon passes) but he found fault with everything and it seemed that it was because John was the proof of Fanny's fall. After hearing my mother's interpretation of how awful Charles had been to my father I was astonished, and I think so was he, when Charles came back from Canada at the age of seventy-three and stayed with us for a few days. He was the sweetest old man, who spent hours with me in the Barn park while I practised sprinting over a measured 100 yards. I was there, in the hall, when he arrived at Little Ardo. He came in soberly, greeted us and noticed the Bodachra grandfather clock standing on the left of the door. Without a word or any hurry he went over to the clock and was disappointed to find that the whisky bottle Old Bodachra had kept there was gone. John Allan was furious with himself for not thinking to put one there for the visit.

John, my father, thought Charles was unkind to him because he was jealous of the affection lavished on him by the old people. Chae, as he was known, was being groomed for hard work and responsibility while John was at the cuddly stage. John had also been the reason for Fanny getting away to the Dominion of Canada and he, Charles, was just dying to be off too but was feeling desperately trapped by his father's encroaching enfeeblement. He was also jealous that, while the old man was impatient with his son's attempts to follow in his footsteps on the farm and had little energy left to help him, he had all the patience in the world for the little boy with the enquiring mind, and liked to take him to look once again at his fields.

If he hated his Uncle Charles, and he did, John loved his Aunt Susie. She protected him from her brother and one of the last things John wrote was a valedictory letter of thanks and admiration to her. And no wonder.

John Allan was a rising star of broadcasting when broadcasting was just rising itself in the 1930s; he started to broadcast to the Empire. The first time he did so he came home to his wife in tears. 'I was wondering the whole time if my mother was listening.' She might well have been, for many exiled Scots, who hadn't been able to wait to get away to the new world and on no account wanted to return, did listen in. But if she was listening she didn't say. When he got into the University of Aberdeen, when he got his honours degree in English Literature, when he became a published author and won literary acclaim for *Farmer's Boy*, Susie wrote and told her sister. She sent Fanny all John Allan's books. But the nearest I ever heard him come to criticising his mother was when he said, sadly, 'I expect she threw them in the fire.' Usually it was a defensive, 'She was just a lassie.' But there was never a sign from his mother. Not as long as she lived. He didn't bore us with it but the space was there.

And those sins of the parents were visited on my father even until he had produced a third generation of his own. It was when, fifty years after its original publication, Century Publishing brought out what was a seventh edition of *Farmer's Boy* in their Scottish Classics series. That should have been a nice honour for the old man as he sank, in his seventies, below the parapet. But his parents' sin caught up with him yet again and spoiled even that. The introduction to that volume starts 'John R. Allan was the illegitimate son of . . . ' He wasn't furious. He was devastated. He put down the volume that he had been looking forward to seeing and said quietly to my mother, 'I don't want to have it in the house.'

My father's Uncle Tom Allan was something of a black sheep. What that meant I was never quite sure, but at any rate he got off to Canada quite a bit earlier than the others. There he seems to have done reasonably well, especially with the ladies. He married one and when he got tired of her or she got tired of him he went through a tribal ceremony with 'a native American' though he called her 'an Indian Princess'. When that fell through he married another white lady. He was always painted in rather favourable terms to me by my father. Tom had been a warm and friendly man and John Allan had had enough of morality long before I was even born. When the old man died Charles and his younger brother Sandy both took the first boat for Canada to join Fanny and Tom, and that was the end of the Allans at Bodachra.

That left John Allan at Gordons College and staying in a little fisherman's cottage at the Bridge of Don with the granny who had been so good to him for all those years. His Auntie Susie, who had tried to protect him, had married Charles Rennie and was busy raising a family of her own. The Allans had made no fortune from their hundred years in Bodachra but there was a little from the roup and his granny told John Allan that, if the money lasted, he would get to the university. It did last, though only just. It is a shame that the old lady didn't live to see him graduate but if she had, the Bodachra money might not have been enough to pay his fare to Glasgow to take up his first job as reporter with the *Glasgow Herald*. With that rail fare paid, the money was all gone and the last connection of the Allans with Bodachra, in the parish of Old Machar, was severed.

So when John Robertson Allan returned from the war to be the farmer of Little Ardo he took over from the Mackies the farm which had been, for remembered time, the farm of the Yulls. He brought with him a new name to the farm, the grandfather clock from Bodachra and a tobacco jar inscribed with 'Mr Allan Bodachra 1893' – but little else of his own family.

11
The Hero's Return

I don't remember actively missing my father when he was away at the war, but there was no doubt that it was a day of great excitement when he was to come home. My mother even made a roly-poly pudding, which she said was his favourite. Apart from a very occasional trifle if someone of great importance was coming, or a tin of fruit with jam and a pourie of cream on top, or her own custard made with eggs rather than custard powder, it was the only pudding my mother ever did make so it added a sense of occasion. I remember that feeling well but I don't remember the spontaneous outburst of joy of which my mother tried to remind me for forty years. Apparently, as I watched her make her one and only pudding ever, I clapped my hands and whooped, 'When my daddie comes home we'll hug him and kiss him and give him all the pudding.' Perhaps we should have, for the pudding was, at least for me, a big disappointment . . . though given my mother's culinary record it should not have been.

The pudding well on its way to being made, I was sent to play down the 'Big Brae', as we called the steep road that led to the Waterside and the village. That was the way Captain John R. Allan would surely pass when he got off the bus that would bring him from Aberdeen. About a third of the way down the brae lies the Breem Hilly, which is just that: a grassy bank covered in broom. The Hilly was also called the Postie's Seat. Though I never saw a postie breaking his climb up to the farm it would have been ideal for that. The broom gave shelter from the north wind, there were boulders to sit on if the grass was too wet and it had a fine view down the valley to the River Ythan and across to Methlick and the Thanage of Formartine. Upstream lie the Craigs, the Braes of Gight and the castle where Byron's mother was born. Downstream lie the woods of Haddo whose more successful lairds, the Gordons of Kelly, swallowed up the lands of the Gordons of Gight in the eighteenth century. On the horizon away to the west lie the Glens of Foudland and to the south Aberdeenshire's blue mountain, Bennachie, and no more than seven miles away to the south-east, the Prop of Ythsie, which was still for me the centre of the universe.

There is no doubt that, for both my parents, my father's demobilisation after four years of short leaves and all the other tensions of a marriage in wartime, was a day of great expectation and joy. I know of no couple who are better suited to each other than they were. They were both socialists. They both preferred each other's company to the company of others. They were both very interested in everything, especially education, and they spent most of their days in educating themselves. They read everything from paperbacks, which they called 'dirt', to politics, philosophy, history, art, architecture and the theatre. They were both graduates of English Literature from the University of Aberdeen and any addition to Eng. Lit. soon found its way into their house. John Allan did a great deal of book-reviewing, among the rewards of which was to keep the book in question. When he died in 1986, I was left with those of his books which had not been stolen and the request to give them to the University of Aberdeen after I had taken all that I wanted. The university got some 7,000 volumes.

Apart from my mother making a pudding, I was not very much touched by the excitement and Albert (the grieve's youngest and my best pal now that he had got over his kicking) and I were busily digging for arnuts on the Breem Hilly when I looked up and saw the khaki figure coming striding along the Waterside from the village. I definitely just saw him, I was not looking anxiously or excitedly. I just saw him. I remember thinking with pleasure but with no hysteria that I knew who that would be.

I have often seen attempts, and so have you, of actors to portray that sort of meeting. In the Hollywood interpretation of such events my father would have thrown his kitbag to the ground, shrieked out of him and started sprinting towards me. I on the other hand would have run for all my little legs were worth towards him shouting, 'Daddy, Daddy you've come home and Mummy's made a pudding.' When we met and somehow avoided injury in a twenty-five-mile-an-hour head-on collision, my father would have swept me off my feet and attempted to squeeze the life out of me while making double sure by smothering me with kisses.

Had my father's homecoming been like that I would have been very afraid indeed. But John Allan, a most loving and gentle man, wasn't like that and neither was our meeting. He just came striding up the steep hill without hesitation, strong and sure with his kitbag over his right shoulder. As he approached me he slowed down and looked at me with warmth but no heat, and said, 'Aye, aye.' He then held out his left hand. I took the hand and we walked up the hill to where the soldier's wife was waiting. I remember my feeling exactly. It was 'look what I have found on the brae'. I knew she would be pleased and I wouldn't mind at all even if he were to get all the pudding.

12
A Long War

It was fashionable at the tail end of the war, and for a long time after it, for people to say they had had 'a good war'. I met a lady in the Bahamas in 1965 whose eyes shone when she told me she had had 'a wonderful war'. And while it had been a horrendous time in many spectacular ways, there is no doubt that a period of national service took a lot of people away from boring ruts, let many see a bit of the world who would otherwise never have been far from the village of their birth. Many mere boys went off to the war and came back considerable men. I still see Lord Mackie of Benshie described as having had 'a very good war', or even an excellent one . . . he had run up that famous overdraft in the pursuit of pleasure, got the DSO and DFC and married the most beautiful woman in the North-east of Scotland, after all. But when my father was asked what sort of a war he had had he answered with one word: 'Long'.

John R. Allan was not a natural combatant. He never once lifted his hand to me or to anyone else that I saw and I was in my late teens and he was unwell when I first saw him show any sign of annoyance. Despite a passionate belief in the government's case, he had not gone with his friends to fight in the Spanish Civil War. He had friends who were pacifists, like the writer Fred Urquhart, and the revolutionising agriculturalist, Joe Duncan, who became the nearest thing he ever had to a father and was much reviled during the First World War for his support of the 'Peace By Negotiation' movement. So it would have surprised no one if my father had been a conscientious objector. But while he didn't believe in corporal punishment and was very much against the death penalty, John Allan was not a pacifist. He believed in the concept of 'the just war' and, unlike World War I, the war against Hitler was it.

What follows is not my best efforts to research my father's war or to produce a document of historical significance. This is just what filtered down to me, his son, from John Allan's own mouth. To the extent that I have remembered it correctly it reflects how he wanted his war thought of.

Despite being a man of letters and already the author of three books including *Farmer's Boy*, John Allan's point of entry was the Royal Engineers, and the step was the bottom one. That meant my father was a sapper, which he always said was infinitely higher than the starting rank in any other regiment. A sapper was much superior to a mere private. To his surprise the square-bashing with which his military training started, while it was hardly stimulating, was no problem and soon passed. Then came the training in the craft of the sapper, which was aimed at providing all sorts of physical back-up to the combatant troops. They had to build gun emplacements, stores and dumps and, crucially, the high-drop latrines so vital to the health of large concentrations of men in the field. Much of that work involved digging, and digging to exact specifications, and in this Sapper Allan showed himself unexpectedly adept. He could dig a hole fourteen feet by three and six feet deep with sheer sides and right angles everywhere.

Sapper Allan had soon reached the standard at which he might have been shipped out to dig holes in the face of the enemy. But at thirty-six he was also reckoned to be a bit on the old side for that and, as he was so good at the job, he was held back to be an instructor. My father was set to spend the rest of the war on Ilkley Moor or on the Downs around Aldershot and Andover, teaching young sappers to dig holes and fill them up again. But the job wasn't all craftsmanship. John Allan had to take what he regarded as an unexpectedly active part in the physical preparation for combat of the troops in his charge. A great deal of time was spent on route marches and endless challenges to get to the top of the nearest and sometimes the farthest hill. As a point of honour he felt it necessary to his standing among the recruits that Corporal Allan be first up the hill on every occasion.

As his son, I am very proud of that fact, though I have always been astonished that he should be able to do it and, even more so, that he should be able to boast about it. John Allan had been a good tennis player at Aberdeen University and he was keen enough to buy himself a set of golf clubs when he had no other of the trappings of even modest wealth. But it says an enormous amount for the power of the army that they managed to get such a natural non-competitor to care a damn who got first to the top of the hill. For the rest of his life he would have argued that, as only one person could be first to the top of the hill, it was an unrealistic goal to set before the young, or indeed anyone.

Though he never said so in my hearing, I think the war was a considerable torment for my father. It took him away from his family, of course, but it also took him away from his books and his career as a writer, which was getting into a rewarding swing at the beginning of the war. My mother told me often of the books that were lost (I somehow imagined them being bombed) as a result of the war. But there were other privations. As an 'other rank' in the Royal Engineers, even when he was Corporal Allan, his pay came to well under £2 a week and, while that would have been little enough had he had it all to himself, as a married man with a wife and child to support, the army sent all but sixpence a day home. That left the Hero with a very small budget with which to indulge his other hobby, the drink. You may well believe that to have been a good thing but he certainly didn't think so. Necessity did mother some helpful inventions, however. John Allan had two streetwise friends in the Engineers and they were able to teach him the gentle art of 'drinking off the piano'. The sixpence a day was sufficient to buy a few fags and a half pint of watery beer and the strategy was to make that half pint last all evening. The tactic was to wait until the beer was flowing and then suggest a sing-song. Wee Sammy from Glasgow (I never got his real name) could play the old piano as well as its quality merited and Willie the Dundee butcher could sing a bit. The piano drinkers would move over to the piano in all the strength they could muster, put their half pints on the piano and, with a great show of camaraderie and back-slapping, get a crowd of drinkers to do likewise. John Allan had a selection of the most exquisite verses of the Ball of Kirriemuir which could be rendered on occasion to get the party really moving. He had composed a verse for each of the characters in the barracks. Just about the cleanest one I can remember went:

A crafter chiel caed Charlie Broon,
Cam ower the hill frae Clatt.
Said 'tho I have a little pin
I like a muckle twat'.

When he first heard that, Sergeant Brown is said to have sent for a measuring tape.

The piano would soon be covered in pints and even the odd short (and not so odd if the Yanks were in for they had plenty of money and liked to show it off), as well as the three half pints of the heroes. From then on it was just a question of drinking what you fancied and fancied you would get away with drinking.

Some twelve years after the war my father was asked at the Ellon British Legion Dinner, at which he was guest of honour, which battle he regarded as the most important in the war. After a quick consideration of the claims of El Alamein, the Battle of the Bulge, Iwo Jima and the Battle of the Atlantic, he startled his audience by saying that, despite the claims of those great battles, he had found the Battle of Andover, if not the most important, certainly the most memorable.

The three heroes were sitting a little disconsolately in the Swan Inn after a hard day's work and in hopes of a hard evening's drinking off the piano. Things were not looking promising as there was hardly anyone else there. Then in walked three enormous American airmen. With their tight breeks and tight fists they swaggered up to the bar. 'Three large scatches, barman.'

The three sixpence-a-day Royal Engineer corporals watched sourly until, six large 'scatches' later, the Americans turned from the bar to face them. One addressed the three pals thus, 'You guys don't seem to be doin' much drinkin.' That's only cat's piss in your glass anyways. Why don't you try a real man's drink?' Holding up his own glass of amber goodness. 'Course you don't get no pay in your man's army. I guess you don't rate for no pay – you caint fight.'

Now the speaker had been right in every detail until the very last phrase. And he might have got one out of three even for that as, up to that time, my father's aggressive ability had been entirely untried. But Willie had defended himself successfully in Dundee's Hawkhill for twenty-five years and Sam had survived all his life in the parts of Glasgow made famous all over the world for fighting. There was no nonsense with them; no squaring up, or saying 'whatwasatyesaid' or being cleverly hurtful. It was sudden and it was straight in.

The adversaries had made a rare contrast. The all-American boys, boyish and clean-cut, tall and immaculate in those neat-fitting, snake-hipped uniforms that were such a hit with the virgins of England during the war. They faced three men who were older in years and much older in experience, while the home team, in their baggy khakis and army boots, were at what might have seemed to be a physical disadvantage; my father at five foot eight and a half inches was considerably the biggest of the three Scots. But they had the traditional Pictish figure, long bodies and short, strong legs which made for speed over short distances.

There would be none of that nonsense proposed for unarmed combat by the Marquis of Queensberry. No straight left to the chin designed to give the taller Englishmen or Americans an advantage. It would be a case of hitting whatever was at the level of whatever you were hitting it with. And if you hit hard enough the rest would soon come within reach.

The three bodies hardened by racing their juniors to the tops of a thousand English hills on utility rations hit the Americans pampered by their legendary PX like torpedoes at the body of a whale. The first to fall had no idea what he had done to provoke a fight when a head exploded the wind from him as it crushed him against the bar. That brought the all-American head down to within reach of the Scottish fists and the fists brought everything down to a handy height for the boots. The second was trying to button his cap onto his shoulder strap, Hollywood style, when he was hit by a similar frenzy of heads, knees, boots and fists. And the third was making for the door when he was stopped by a chair . . . and not one of those Hollywood things that splinter at the slightest impact, but a solid piece of good English beech. The three Americans, in a state of shocked semi-consciousness but total submission, were then picked up like three sacks of very expensive and well-appointed potatoes, carried across the street and dumped in the duck pond.

There was no need to sing much that night. Though everyone was glad that the Yanks were over here, there was a considerable feeling that they would have been even more welcome had they been a little less oversized, oversexed and overpaid. The heroes of the Battle of Andover were on free beer for a week at the Swan.

It would be nice to think that it was for his part in the Battle of Andover that Corporal Allan was put up for the officer selection course. But he would have nothing to do with the suggestion that it was anything to do with his merits as a fighter. He told me often that it was not because his superiors thought him 'officer material'. On the contrary, as they had proved that he wasn't 'other ranks material', they wanted to make him an officer to hide their mistake.

Another of his excuses seems more likely to have some truth in it. Apparently they wanted him to do some research and writing for propaganda purposes and, as most of the people he would be speaking to would be officers, it would be easier for him to get sense out of them as a brother officer. Also it would save the bother of moving heaven and earth to find neutral ground for their meetings. Far better to give the researcher access to the Officers' Mess, where they could meet in congenial surroundings. Despite his predisposition against officers, my father was a most willing applicant for promotion. The money his pay as a corporal allowed him to send home to my mother and me was certainly not sufficient to keep us in the manner to which we aspired, and he was very fed up of the sixpence a day.

I only know two parts of the officer selection. The first is a solid hour of physical training to make sure the aspiring officers were made of the right stuff. But it turned out that the right stuff was more complicated than many of the younger hopefuls thought. They thought that the way to selection was to do the widest legs astrides and the straightest press-ups, whereas an older head had tipped Corporal Allan off that all that was necessary was to keep something moving for the entire hour. That he did while many of those who started more impressively could not.

The other part of officer selection of which my father told me was a series of interviews designed to probe, in depth, Allan's suitability as officer material. They produced at least one bad moment. He was asked about his hobby of reading. What did he read? That had required a long answer. Who was his favourite author? That too was a long list. It depended on his mood and what he wanted to read at the time. But if he was forced into making a choice he might choose Virginia Woolf.

The questioning officer was visibly shaken by this answer. He asked why Allan liked Woolf and he wouldn't take any sort of reasonable response for an answer. My father thought he had blown it. My mother and I would have to survive on a corporal's pay for the rest of the war. Perhaps in the world of officer selection you were supposed to give tough titles. Scott or Stevenson, of whom he had tired many years before, perhaps, or Hemingway, whom he did admire. Perhaps the selecting officer thought that Virginia Woolf was the sort of choice you'd expect from authors and silly buggers but hardly of officers and gentlemen, however temporary. Eventually John Allan had to ask if the officer had anything against Mrs Woolf. 'No. No, not at all,' came the reply. 'It's just that she is an aunt of mine and I could never make head nor tail of her myself.'

At any rate my father did get his commission and was able to do, for the rest of the war, what he had done in civilian life: he wrote, and, though his war was still 'long', he stuck it out until after VJ Day in 1945, when he was demobbed with two excellent, hard-wearing blue pinstriped suits, a .22 revolver and the rank of captain.

I really should have written 'from the rank of captain', rather than 'with' it as John Allan had the utmost contempt for amateur soldiers who like him had joined up to do their bit but who carried their rank back into civilian life. That was all right for professional soldiers but he thought the temporary gentlemen should revert to plain Mr.

13
Joe Duncan

With neither of his parents around, the new farmer of Little Ardo might have had a rather unbalanced family life. To some extent he did. The Yulls were overseas but the Mackies were still in Scotland and my mother's people did dominate things. They might have anyway for they are interesting, generous and sociable people. But the absence of opposition did make our family unbalanced – with numerous Mackies, no Yulls, only one Allan and no (local) Robertsons. It was only much later that I realised how odd it was that I was proud of the growing number of my first cousins. There were eventually eighteen of them, all Mackies. But I never thought to wonder why I didn't have another eighteen from the Yull side, let alone another eighteen each from the Allans and the Robertsons. Those seven Robertson aunts in California might well have done better had we made contact with them.

So the family was basically the Mackies. But not totally. For John Allan's aunt and protector, Susie, was still around and was undoubtedly his mother figure, though his real mother had left an unfillable hole. And though I saw no sign that John Allan felt the lack of a father other than when he presented his birth certificate with 'illegitimate' scrawled across it, there was a very close friend who was some part of a father figure. He was Joe Duncan, founding secretary of the Scottish Farm Servants' Union. On our trips to Aberdeen in the Austin 10 we used to break our journey home by turns with a stop at North Ythsie, where I got a chance to raid again my grandmother's garden or see if there was anything left in the sweetie dish in the corner cabinet, and a stop at either Auntie Susie's or at the Duncans' where they were always very anxious to talk to my parents and shut me up with a bar of chocolate. One week my mother's people. The next week my father's.

Joe Duncan was a truly remarkable man who deserves a book to himself, and indeed he got one, written by Joe Smith, an agricultural economist at the North of Scotland College of Agriculture. Joe Duncan was a self-taught sma' hudder's son from Banffshire who left school at fifteen but rose to be an examiner for doctoral degrees at Oxford (with no degree of his own until he was awarded an honorary doctorate by Glasgow University) and who had no other means of support than what he earned from his work. The main work for thirty years from 1912 was for the Farm Servants' Union, most of them as secretary. It was said that he knew every farm in Scotland and that he got round them with public transport and his bike to save on expense. But that wasn't anywhere near the half of it. He sat on two Royal Commissions in 1912 and 1919. He served on six Scottish Departmental Committees. He was on the Agricultural Research Council, the Scottish Agricultural Advisory Service and the Scottish Agricultural Wages Board. He was a governor of the Scotland College of Agriculture, retiring from it as its chairman. That is less than half the list but I am fed up of it and want to get on to higher things.

For Joe Duncan, with no formal education beyond Gordon's College or the age of fifteen, was grateful for the help he had received from the Worker's Education Association, so he lectured a good deal for them and ended up chairman of that too. And he was right at the heart of academic economics. Joe was one of the three Scotsmen – the others were Jock Currie and John Maxton (Jimmy, the great Independent Labour Party MP, pacifist and Red Clydesider's brother) – who were responsible for founding the Institute for Agrarian Affairs at Oxford, and was one of the founders of the International Association of Agricultural Economists. He was President of the Agricultural Economics Society and as such was summoned, with L. K. Elmhirst, his chairman, by the permanent secretary for Prime Minister Clement Attlee to explain the opposition by the Research Institute at Oxford to some government plans for silos. The government had favoured one firm and the deal smacked of cronyism, but on the Oxford Institute's costings that firm didn't come top of the list.

This account, by Elmhirst, shows the toughness and incisiveness of the little man from Banffshire. The permanent secretary welcomed the two:

> 'Good morning, Mr Duncan and Mr Elmhirst. My minister is much perturbed over some of the publications that are being issued by the Institute of Agricultural Economics Research. This is as you know a National Institute supported by public funds. The recommendations do not coincide with the Minister's policy.
>
> 'Is it,' Joe asked, 'the accuracy of the assembled facts to which the minister takes exception?'
>
> 'No.'
>
> 'Is it that the conclusions drawn are unfair conclusions?'
>
> 'No, Mr Duncan. But the conclusions go in the teeth of the minister's declared policy and he will not have it.'
>
> 'If the facts,' commented Joe, 'are relevant to the subject, Mr Secretary, and if the conclusions are fairly drawn, may I suggest that, though the minister may not like them, this institute is performing exactly the job for which it was set up.'
>
> Without further ado we both got up and went out.
>
> *The Scottish Farm Servant and British Agriculture*, J. H. Smith

Poor old Sir Humphrey. I wonder how he put that to Attlee?

Joe's best work may have been in the field of international agricultural trades unionism. He was founder member of the International Land Workers Federation and a regular at the International Labour Office in Geneva. He was one of the chief founders and loyal supporters of the Oxford Farming Conference, which still shapes the thinking of politicians, academics and practical farmers in the twenty-first century. He was a giant in his day in Scotland, though he was small and slight. And he had the ability to fill a big hall so that everyone could hear him even if he spoke in a conversational way.

But Joe Duncan was at bottom a trade unionist who dedicated his life to promoting the welfare of the Scottish Farm Servants. His philosophy was that the way to improve the workers' lot was to improve the industry as a whole. It was the duty of the workers to agitate for more wages, better houses and more leisure to force the farmers, who he saw as a conservative lot, to adopt mechanisation by making the price of labour too high for the waste of the old labour-intensive ways. If the farmers could be made to spend on machines, if there could be built modern cottages owned by the local councils rather than cottar houses tied to the farms, the industry would be able to attract and keep better men who in turn would deserve and get better wages.

Joe's ideas for promotion of the farming industry were not understood by many farmers, who thought him the very devil. After the great meeting at Ellon where he had told James Low and the rest of the workers that if they stuck it out for £32 for the half-year and none of them gave in they would all get it, Joe Duncan took the train back to Aberdeen to attend to the rest of his business. The carriage was full of farmers who had spent the day trying to hire men for less than £32 and they didn't recognise the devil sitting in the corner. Their news was all about 'that bugger Duncan' and what they would like to do to him, and how he should be 'in the jile'. And if that bugger Duncan would just disappear they could have got the lads to fee for three pounds less in the half-year.

The farmers all left the train at Bucksburn. Joe was just getting used to all the extra space when he noticed that one of them had left his wallet. Joe ran after the farmer who was very grateful. 'And who can I say the honest man who returned my wallet is?' he asked.

'You can just say your wallet was found by that bugger Duncan,' he replied and returned to the carriage.

Joe Duncan was of course a socialist, but his was a very down-to-earth socialism. Joe wanted to improve the lot of the poor, not to complete any fancy theories. He knew all the great leaders of the labour movement. He knew George Bernard Shaw, Professor R. H. Tawney and the Webbs as well as the politicians like Keir Hardie, the founder of the Independent Labour Party which Joe supported, Jimmy Maxton, Paddy Gallagher and all the Red Clydesiders. He could be very scathing about intellectuals who had no practical idea of how things worked. He called the intellectual Pethick-Lawrences the Pathetic-Lawrences. And he couldn't be doing with tokenism. Although an avid advocate of votes for women, he would have nothing to do with double-barrelled names. When his wife-to-be, Mabel Saunders, said they should maybe call themselves the Saunders-Duncans Joe was totally dismissive: 'What would happen if we Saunders-Duncans had a child who wanted to marry one of the Pathetic-Lawrences?' The question was never raised again and sadly they never had any children, which no doubt encouraged Joe in his role as father-figure to the farmer of Little Ardo.

Joe Duncan's undogmatic socialism was well illustrated in 1932 when he supported and encouraged all his members to attend a great rally at Pittodrie Stadium in Aberdeen. It was at the depth of the Depression; the year when almost every big farm in Aberdeenshire hired at least one fewer men. It was the time when, after the unsuccessful feeing markets and the loss of their tied houses, Inverurie Town Hall was filled with families with nowhere else to go.

There were those who said that Joe was wrong to support the farmers in their rally for higher prices. The farmers agreed to give the workers a day off to go to the rally – and at that time the farm servants worked a six-day week and an eleven-hour day, unless they were horsemen, in which case it was a good hour more. The leader of the National Farmers Union of Scotland at that time was none other than my grandfather, Maitland Mackie. It was widely thought that Joe had sold out to the farmers and that Maitland Mackie had pulled off a masterstroke. The workers had admitted times were difficult for the farmers, thereby allowing that the farmers were right to cut the wages and cut down the staff. Joe saw it differently. He was not a class warrior who saw his struggle as a battle against the bosses. His battle was to improve the lot of the workers by forcing the farmers to adopt more efficient working practices and to force the government to support the industry. Besides that and to that end, by getting Maitland Mackie to agree to let the men off for a rally he had weakened their arguments against the five-and-a-half-day week, which was another of the union's objectives, another way to get the bosses to mechanise by making labour a more expensive commodity.

When first I knew Joe Duncan and Mabel they lived modestly in the unused farmhouse at Tillycorthie Home Farm. Joe was a governor of the College of Agriculture, and Tillycorthie was, as it still is, one of the college farms. The house was filled with the books of this man who had left school at fifteen. They shared the house with Mabel's elder sister and Joe's long-time secretary from his days with the Scottish Farm Servants' Union, Miss Belle Jobson. They had a very old green car which might have been a

Morris 8, and when, in the late 1940s, the four of them crammed into the car for their frequent trips, their combined age including the car was said to be 'gey near 400 years'.

They were a lively lot, although death could not be very far away. They all had stories, one of which was told by Belle herself of the time she had been looking after Ramsay MacDonald, the first British Labour Prime Minister, at a political meeting in Aberdeen. A hostile crowd was waiting for Ramsay MacDonald and there were rumours that the fascists had been down hiring a mob at the docks. They tried to make an escape by the back door of the Music Hall, but there was a crowd of thugs covering that. Belle was most amused when the great man who was to lead Labour to power for the first time said, gallantly, 'You go first, Belle, they maybe winna hurt you, being a woman.' Belle had had a look at them and had her doubts. They survived by adopting Belle's much less ambitious strategy of locking themselves into the Music Hall until the thugs had got thirsty or had all gone home. All the same, she thought the signs were not good for the leadership qualities of the future prime minister.

Joe had worked all his days for the workers of the world and for the farm workers, whose improvement he saw as the way to promote the advancement of farming. Joe's glittering career, which took him all over the world and through the corridors of power and the Halls of Academe, left him with several thousand books, many friends, and many more admirers. But it did not make him rich. Most of his life was spent in the service of the Scottish Farm Servants' Union where even as its General Secretary his salary was £3 a week. At that time Belle was paid more than her boss, as were most of the clerical staff, who had to be paid the rates negotiated by the Clerical Workers Union. When he retired there were but the four pensions, and when his wife's sister died there were only three. When the college wanted his tied house the Duncans somehow got a little semi-detached house called Witchhill in Newburgh, where it backed onto the Ythan estuary. My parents and I used to visit them there. The ladies made a fuss of me but the great man didn't patronise me with any questions about how long school holidays I had left or how the football was going. He had no time for small talk. He had important things to discuss with my parents and political jokes to make. I was welcome but I got my bar of chocolate and went out to play.

The Labour party had tried for many years to get Joe to take a safe seat and become a Member of Parliament, but he wouldn't hear of it. He was far too honest to have managed all the compromises politicians have to make. He stood three times in hopeless seats where he was quite satisfied that he had no fear of being elected, but he would have been miserable at Westminster. The Labour Government tried hard to get him into the Lords in 1945 but he told them to go no further. He was a socialist who believed he was no better than the next (or even the last) man, though that was manifestly untrue, and he would not on any account be put above any man. Among the many honours that were offered he did accept a doctorate from Glasgow University, but that was different. Degrees are for achievement and do not imply anything to do with class.

That degree was what Dr Duncan had to show for all that public work and certainly he had no money. When he was eighty and the old Morris packed up he couldn't afford another car, which by that time would have cost £100 or more. So a group of his admirers got together and put up what they called the Gadabout Fund. My parents were its trustees and it paid for taxis for the Duncans for as long as they were fit to take them.

There were few monetary rewards for all his efforts but there were others which he cherished more. My parents sent the foreman, Bill Taylor, down to Newburgh with a load of sticks for the Duncans' fire. When he arrived the old man told Bill just to coup them in the street and he would take them round to the stickshed at the foot of the garden at the back. He had a small wheelbarrow and he said he would easy manage. Bill Taylor asked to see the logistics of the problem and then insisted on doing the job himself saying, 'Efter aa that ye've deen for his [us] I could surely shift a puckle sticks tae ye.'

14
Captain Allan and the Election of 1945

There was a lot more to do than just eat up the roly-poly pudding when the Hero had climbed with his son the half-mile up the Big Brae to his new home. He had to re-establish his relationship with his wife and to establish one with his son. That meant more than just getting me to stop banging on about how much I admired private enterprise. I must say he followed his splendidly easy-going meeting with me on the brae with a masterclass in how to make life interesting and fun. Mostly that meant a hands-off approach – give the lad room. But he was first-rate on the ballads for which the North-east of Scotland is literally second to none in the world. More traditional ballads were collected within twenty miles of Little Ardo than anywhere else in the English-speaking world. Our music room was the Austin 10 in which he toured me round Buchan and Formartine explaining the way of things and singing. He sang me 'The Bonnie Lass o Fyvie', 'Mormond Braes' and 'Muckin o Geordie's Byre'.

But my favourite was of John Allan's own making. To the tune of 'McNamara's Band' he sang:

Oh we drive about the country in our forty horsepower car.
We don't care where we're bound for – be it near or far.
The engine roars and the engine snores and the horn it gangs 'toot, toot'
And the Bobbies aa blaw their whistles and cry 'Oh, oh my God, look oot.'

That was a real winner, especially when the passenger was allowed to do the 'toot, toot' on the car's horn. I loved that song, as did my children and as, in 2008, do the composer's great-grandchildren.

So John Allan made a good job of getting on with the son who hadn't noticed if he missed him, but he also had to get the farm reorganised in recognition of the transition from one of Maitland Mackie's outfarms to the home farm of Jean and John Allan, for they were to be partners.

But before we come to that, in common with what turned out to be the overwhelming majority of the British people, the new farmer of Little Ardo saw the ending of the war as a chance for a new beginning in Britain. There should be no return to the hard times between the wars. For that, a new government was necessary. And John R. Allan was an obvious choice as a candidate for Labour.

He was a brilliant speaker who had carried all before him in his university debates in the 1920s. He was so popular there that in one serious political debate he had to be brought back from the bar to do an encore because the applause wouldn't die down sufficiently to let the debate continue. Later, he gave what was believed by many to be the finest after-dinner speech ever made, perhaps because it was the shortest. It was at the Aberdeen University Old Boys And Girls Club, the Elphinstone Society's Glasgow branch, where he was down to propose the toast to the university's 1495 founder, Bishop Elphinstone. He stood up and said, 'My good friend the Bishop Elphinstone and I . . . and I . . . and I . . . ' and slid gently under the table to thunderous applause. Which of us has not longed for an honoured guest to do that? He was a popular and prolific broadcaster on the radio, which would help counteract Bob Boothby's national fame as both former minister of the Crown and long-term MP for the constituency of East Aberdeenshire.

It was thought by others that John Allan had had a good war. He was a 'ranker', one who had entered at the bottom and risen through the class barrier to become an officer. That sort of thing goes down well in Aberdeenshire. He was a local literary figure of some importance. He was a very well-known broadcaster and had just come to be a farmer in Buchan. He was the ideal candidate to stand against the great Lord Boothby (Bob as he was then). It was not what my father would have chosen to do on his return but he was willing to do his bit. After all, there wasn't the slightest danger of his unseating the wonderfully popular Bob Boothby. My father, with some reluctance, allowed his name to go forward and with a deep sense of shame, he even allowed himself to be billed as 'Captain Allan'. He regretted that for all of the twenty years or so it took him to lose his unwanted title.

I remember the excitement of the campaign well, though I was never taken to any of the hustings. It had been a very lively campaign – both candidates visited all the fish and cattle markets and spoke in every single village hall, sometimes four speeches and three dashes to the next village hall, in a night. After that, for the Labour candidate, it was back to Little Ardo for smoking, drinking and laughter about the disasters of the day – like the tale of the earnest young canvasser who had been visiting farm cottar houses and telling the wives of all the benefits socialism would bring them. She had been telling this tractorman's wife that the scandal of tied cottages would be ended by a Labour government, which would build agricultural workers' houses so the workers could be independent. The young canvasser had been waxing eloquent about how these council houses would have at least three rooms and, the coup de grâce, flushing lavatories. Unfortunately she was canvassing at the fine modern farm of Westertown of Rothienorman, my uncle Mike Mackie's place, and just at that moment she heard from inside the cottar house the unmistakable sound of a water closet being emptied.

One of the essentials in these marathon evenings of meetings was someone to hold the fort until the candidate, who always ran late except when he ran very late, arrived. The meetings were scheduled to start at forty-minute intervals, but with interest in politics at an all-time high, and the villages up to ten miles apart and the Labour Party's cars being below average, the chairman often had a difficult time. John Allan cherished the memory of running very late and the performance of a young lass who had to keep the good folks of Foggieloan happy as the candidate fell further and further behind. The chairman was, like so many activists in the 1945 election, new to politics, having been drawn in by the fervent (rather than informed) desire that things should be better for the working class after the war. Now Foggie has always had a slightly tough reputation and the candidate was anxious as he drew near to the village hall over an hour behind schedule, but he need not have worried. His young chairman had told the voters all she knew about socialism and about their candidate. Then she started to sing to them. She had a fine voice, a good sense of time and she knew all the auld Scots songs. By the time the candidate arrived the village hall at Foggieloan was rocking.

It was not all folksy anecdotes. I remember my mother incandescent with fury at Boothby. He had been asked a question about his leader, Winston Churchill. Notwithstanding the Old Bulldog's stunning success as a war prime minister, were his credentials as a prime minister for peacetime not suspect? Reference was made to his performance as Chancellor of the Exchequer, when he had certainly made the inter-war years worse by his attempt to preserve the Gold Standard. Boothby had answered dismissively, saying, 'Churchill knows nothing about economics'. The Labour candidate's wife thought that was disgraceful and probably true. I remember her saying with a strange wild look in her eyes, 'Economics is about producing food, employment, housing and health. What use is a prime minister who knows nothing of economics going to be to us?'

My parents both knew Boothby quite well and liked him very much. They believed him to be a socialist at heart and he certainly did agree with much of what the Labour Party wanted. So my mother asked him when he was going to join the Labour Party. 'Not till my own party is down and out.'

It was the first of many excursions by Maitland and Mary's Mackie's next generation into national politics and it was uncomfortable for them. They had voted Conservative all their lives and Boothby was Mary's political idol despite his liking for a droppie. She was a Tory, though Maitland was more of a pragmatist. He wanted to see the job done and done well and didn't mind too much how it was done. Don't forget, it was he who converted his eldest daughter Jean, John Allan's wife, to socialism. At any rate, on election day the old man was unusually quiet. He was clearly wrestling with a problem. Then at lunchtime he announced that in the afternoon he was going up to Tarves to vote for John Allan. Granny was shocked and scandalised. She ridiculed the suggestion that her husband would vote for his son-in-law and forbade it. One vote for each of the two candidates didn't seem worth going to Tarves for and certainly wasn't worth fighting over. So they both stayed at home, forgoing their democratic rights for the first time and in what may have been the most important election in the nation's history.

The count was held in the Peterhead Town Hall and it was there that Captain Allan enjoyed anxiety far worse than anything he had encountered in his long war. When the counting started he immediately went into the lead. As the night wore on the lead grew. Surely after four years away and looking forward to nothing so much as being home, surely he wasn't to be condemned to another four years in London? He had agreed to be a candidate in a hopeless seat just to show the red flag and in the certain knowledge that there was no danger of election – and yet here he was with a commanding lead.

As the night wore on it became clear that it was going to be a Labour government and that things were never going to be the same again. Ernie Bevin would soon be saying, 'We are the masters now', and it looked like the farmer of Little Ardo would be one of them.

But something odd was happening. Though Labour was sweeping to power in England and Wales, in Scotland the Tories were hanging on. My mother said it was because Mr Churchill didn't campaign in Scotland, but for whatever reason, the swing to Labour was muted north of the border. The count at Peterhead which had given my father such a fright had flattered and deceived him. The early votes had been from the towns of Peterhead and Fraserburgh, where Labour had always been strong. But when the rural votes came in and were counted the next day, all was put to rights. Mr Boothby was returned, though by fewer than 2,000 votes. Captain Allan was delighted. The country had got the government he thought it needed, he had done his bit with honour and he didn't have to be part of it.

15

Post-war Reconstruction at Little Ardo

The first thing that had to be tackled at Little Ardo was the house. The grieve had the Victorian extension at the back and until we arrived, the Georgian half had been tenanted by three merry, in the sense that they laughed a lot, ministers' widows. What was good enough for the widows wasn't good enough for us. For all that Little Ardo had had the first water closet in Aberdeenshire, when I arrived we only had a thunderbox in what I am pleased to say is now the little garden sheddie. Because of the way the house was divided, the grieve had the water closet.

So the profit from the sale of Croft House, the house to which I was born in Blairlogie, was invested in a new extension. At the tail of the war there was an enormous queue for building materials as the new government set about building to stave off what everyone assumed would be a post-war depression similar to that after the First World War and everyone set about rebuilding after the bombs and investing for a slice of what was to prove to be the longest-running boom in British history. The key to getting contracts was showing that you could supply the materials and that, rather than his skill as a mason, my parents always said, is why Jake Smith got the job of adding the latest Georgian wing to Little Ardo farmhouse.

All right, it was only the reign of George the Sixth, but we had hoped for better. The walls soon cracked and whenever the north wind blew the water poured into the kitchen. The slates Jake had access to were just too short for the pitch of roof he put up. I mustn't give the impression that Jake was unloved by my family. Indeed he was a bit of a hero. Apart from the fact that he took on the job when no one else had the time or the materials, I don't really know why. He certainly was a hard worker and didn't like to waste time. If either of my parents went to discuss progress with their master-mason, he would take the opportunity to make his water. Jake wore a slater's leather nail bag which hung down like a sporran at a handy height. My mother was amused by the stream of water that always emerged from behind the nail bag as they discussed progress.

But it did give us a kitchen which at that time was able to pass for modern, and it gave the grieve and his family of five a much-needed extra bedroom and a bathroom. That left a spare bit right in the middle. A tiny room with no external wall, no ventilation, no power points and no running water. From the garden sheddie my mother, a student of architecture who could tell you all about the design of the Coliseum with a profound understanding of design for living, introduced the Elsan. This earned from my better-off cousins the name of the Dirty Bathroom, though, if the disinfectant was anything like as strong as it smelt, the little room was anything but unhygienic.

It had been quite pleasant in all but the worst weather to be out in the garden sheddie doing your business. It only had half a door, there being maybe six inches missing from the bottom and two of the five feet or so missing from the top. Rather than contemplate a bare wall you could see the world go by. I remember being visited by a robin. But when the Elsan was moved inside, not only was there nothing to look at but there was nowhere for the stink of Jeyes fluid to go except up your nose or into the house. By the end of 1946 Jake Smith had got the Dirty Bathroom done away with and a WC installed and upstairs one of the four bedrooms was converted to a three-piece bathroom.

Though James Low had pronounced the steading done when he arrived to be grieve in 1930, it was still done but still in service in 1945. Built in the traditional fashion around the midden, the original building to the north had been a single byre built at the end of the eighteenth century with stalls for twenty cattle. Then when my great ancestor William Yull came to the farm in 1837 or thereabouts, he built what had been at that time a great barn on the west side. It stood all of twelve feet to the eaves and had a built-in thrashing mill which gained its own notoriety.

It was made by Geordie Paterson, a famous millwright from Backhill of Ardo Croft. It was designed on the principle of continuous flow. The sheaves would be forked onto the table from carts which would be outside and the grain and straw would land on the inside, from where it would be stored in the barn while the grain would be carried up the stairs to the loft. All went to plan and there was no doubting that it was a wonderful asset. Sadly, though the concept was very good, the plan was inside out. Geordie built the mill the wrong way round. So every sheaf had to be manhandled into the barn and then forked up onto the table from which it was fed into the mill. And all the straw and grain landed outside and had to be manhandled back in again. Seldom can a labour-saving idea have caused more work. The only plus in the exercise was the pleasure it gave to the neighbours, who were pleased of the break from Little Ardo being first with everything and not afraid to say so.

By 1945 the Old Barn had become a cattle court because Maitland Mackie had by this time built a grotesque twenty-foot high cement-brick barn stuck onto the north end of the old byre. That housed the new thrashing mill and kept enough straw for the stone-built double byre, which he also built in the 1930s. It stands on the east side of the midden and was Maitland Mackie's great addition to the Little Ardo steading. It became the main shed, as it housed the cows, which provided Little Ardo's main cash crop – milk. The square around the midden was completed on the south side by a two-pair stable which, since the horses had been replaced by tractors, was a store for anything that could go into a door four feet wide. Next door was the tractor shed which, with some expert and patient driving, could be made to house three tractors. Then there was the shoppie, where the hand tools and workbench were kept, and the glass house, which Maitland Mackie had built as an incubator house for hen's eggs. But now it was reduced to being a store for unwanted bits of second-hand binder twine and used straw rapes and the baskets into which the potatoes were gathered and which, for some reason, were called skulls.

All that was on the north side of the close, which was bounded on the south side by two of the ugliest black sheds I have ever seen. They had been built of wood and creosoted to house several hundred hens each. By the time the Allans arrived there were no hens, so one of the sheds was demolished to give the grieve a kitchen garden, now that the farmer would need the original one. The other was converted into two cart sheds, a stick shed and one which was a haven for rabbits, guinea pigs, cats and occasionally Mrs Low's hens.

At the east end of the close there is a handsome, freestanding, stone-built 'tattie shed'. It wasn't built for potatoes but to house the itinerant workers who used to come from Aberdeen in October to gather the twenty acres of potatoes which Little Ardo then grew. The tinks (as they were known without political correctness, affection, rancour or contempt) had slept there and cooked for themselves before the war, but by the time we arrived at Little Ardo the tubers were normally lifted by a squad of schoolchildren. Sometimes those came daily from Aberdeen and sometimes they were from Methlick School. The tattie shed was used to store the seed potatoes, once they had been dressed into their boxes ready for sowing in April.

That was a magical place for boys in winter. The enemy of the seed potato is frost. Once they were dressed into their trays they were very vulnerable and a brazier had to be kept burning day and night in the tattie shed. Sadly that attracted the seed potato's second enemy – the small boy. It was attractive because it was fine and warm but, even better, we could steal the seed and roast them in the ashes that lay beneath the brazier. If you could 'borrow' a wee bit of butter, with salt from the cows' licks, they made a culinary treat beyond compare.

The tattie shed wasn't nearly big enough to hold the bulk of the crop, which was grown to supply seed to the English market, but luckily there was another shed converted for that job. The piggery stood apart from the rest of the steading down the brae and to the west. It had been built before the war by Maitland Mackie at a cost of £300 and though the pigs were all gone the piggery was a wonder. Every lorry driver who landed at Little Ardo stood amazed at the sheer size of it. I can clearly remember one driver who was not satisfied when told it was 152 feet long by 46 feet wide. He insisted on pacing it out. It was only seven feet at the eaves so that, with the increase in the size of farm machinery by the time I was the farmer of Little Ardo, it had become a miracle of inconvenience. But at the tail-end of the war it was a great asset for wintering the potatoes and for storing any bits of old anything you couldn't bear to throw out.

The only immediate improvement to the steading deemed necessary when the Hero returned from the war was the cementing of the close. And how well that was done. Sixty years later it is still there and doing its job. It meant that the dubbs was never more than an inch or two deep, except when the men had scraped it and swept it into heaps. It seemed that there was never quite enough time to dispose of the heaps, so the scrapings and sweepings were left to find their own level.

Standing as it did to the north of the farmhouse, the steading did afford us some shelter from the worst of the weather, though the prevailing wind came howling in from the west down Ythanvale from Fyvie. Sitting on the brow of its little hill, Little Ardo was a windy place. The obvious solution was trees, but we in Buchan have no tradition of planting trees. We tend to regard trees as vermin, like sheep, a waste of good grazing. But Captain Allan and his lady had both been brought up in the doucer environment of Formartine, where personal comfort counted for something, so an ambitious programme of tree-planting was undertaken.

Of course, even with a staff of six there could be no question of the workers wasting time on such frivolous pursuits. If the farmer wanted trees he could plant them himself. He made a start on a bleak November day in 1945. I should have said 'we' made a start, for the farmer was not altogether alone. My role was to bring each tree after my father had dug the hole and then to hold it as upright as I could while he filled the hole and firmed the earth with his feet. I was very pleased to be helping with the planting because as far as I was concerned, it was all being done for my benefit. I had complained often to my mother in the early days at Little Ardo that, as there were only thirteen great big trees and no decent hedge, there were hardly any birds' nests worth looking for and that therefore Little Ardo could never be a match for North Ythsie.

The plan was for a mixture of Japanese Larch and Scots Firs to grow quickly and offer us as soon as possible some protection from the wind. Then there were to be a few slow-growing beech and copper beech trees to give us a bit of majesty, which my father had no chance of living to enjoy but which he told me I just might, though even I would never see them full-grown. It takes real love of the land to plant for another to reap, but John R. Allan was pleased to do it and I am pleased that I was there with him. There was one beech tree, the tallest of all at five feet, which I remember planting as clearly as I can remember anything. My father was a bit doubtful about planting it at all because its leading branch had been broken and it was being led by forked stems. Sixty years later it is still there and I gaze at it in wonder. It is still split-stemmed, in fact it now seems to be split into four, and all tower above the house. I can no longer get my arms round my tree but I fear I would have to live another hundred years to see its full majesty. I also fear that this greatest of our trees may one day be felled because we planted it too near to the house. But I wouldn't like to live to see that day.

I don't know what James Low thought of the Hero's return, but it would be surprising if he had been entirely happy about it. He had been very proud of running one of Maitland Mackie's farms for him and the prospect of having a boss who could not be in Mackie's class constantly at his elbow cannot have been a pleasing one. In one respect he would have been pleased, however. The arrival of a farmer about the place seemed to offer another pair of hands at harvest.

My father had not long returned when he found himself forking on the land, that is, using a two-pronged fork to put sheaves from the stooks in which they had dried into a cart for transportation to the cornyard. There they would be built into stacks which would keep them dry through the winter. It was one of those rare harvest days when the sun beat down out of a clear sky. As an extra bonus, a drying breeze made the oat sheaves 'reeshle', which means something like rustle. Three carts were in use so at any one time two would be in the cornyard being unloaded by the tractor-men, a sheaf at a time, as James Low and Bob Gray built a stack apiece. On the land the third tractorman was driving his cart between the stooks while my father and the second tractorman's wife, Mrs Glennie, forked the sheaves onto the cart, where the hairst-man built the load.

It all looked fine to me as I sat on the dyke at the front of the house looking at the goings-on in the mysteriously named Lotties field. But what did I know about it? James Low wasn't pleased. Fine days in harvest were not for enjoying. They were for catching up. The load at his ruck was empty and the 'forkers on the land' still hadn't even loaded the next cart. He slid down off the ruck and charged, without actually running, across the field and grabbed the farmer's fork out of his hand and started putting the sheaves up in threes and fours. I swear there were several stooks (eight sheaves to the stook) of which Mrs Glennie didn't even get to put one sheaf into the cart. The hairst-man, whose job it was to build the sheaves into a tidy and stable load, was about buried in a tidal wave of sheaves. The tractorman didn't even have to stop at the stooks. Very soon the cart was loaded and Jimmy slammed the fork into the stubble at his employer's feet and said with wild eyes, 'Noo, that's foo we fork sheaves at Little Ardo.'

And that was the last time my father did a hand's turn on the farm. He kept the books and the profit, when there was any, and left the work and most of the organisation to James Low. That gave the grieve the chance to farm and the farmer the time to write. It was a most rewarding partnership. The land was kept in good heart, and the journalist who had come courting Miss Jean Mackie at North Ythsie was able to do his best work without developing callouses on his hands.

16
James Low

James Low was a very important figure in my little world. He provided the hardness that my parents lacked. I don't think that he ever gave me the thrashing that was often deserved but the real threat of it meant that it was not deserved as often as it might have been. He knew the right way to do everything. 'There's only ae way tae dae a thing at Little Ardo and that's the richt way.' His contempt for anyone who did anything badly was awesome. He knew all about body language before clever people gave it a name. It was James Low who gave me the best insight into what being a farm servant in the hard times was like. He told me the old stories – like the loon who was hand-loading a cart with neeps. The model was you had to pick up and throw two turnips at a time. A good man aimed to have six neeps moving all day. He had to have two in his hands, two sailing through the air while two were settling in the cart. When the neeps were frozen, as they often were, for they were winter fodder, it was just about as cold a job as your hands could get. So this loon was complaining to the grieve that the neeps were cold. The unsympathetic reply was, 'Well dinna haud them sae lang.' Robbie Rattray, one of the Little Ardo loons, told John Allan that he didn't want to work on the land when he grew up. 'It's ower caul on the hands puin neeps. I wint tae work in a shop and be a grocer.' I hope he made it.

James Low told me about being a ploughman and having to wrestle with the plough on daily hikes behind his pair for up to twenty miles. He told me about the terrible chaffing between the legs caused by all that walking and all that sweating. He also told me what they used to do about it, though I have never been convinced that this was not just another piece of folklore invented by those who have lived through great times to dazzle those who have not. Among our marbles were little clay things called 'pizzers' which had hardly any value. Then there were sparkling 'glassies' which looked like sooking sweeties. And then there were 'dollars', which were perhaps an inch across and were worth a few glassies or many pizzers. Now, according to Mr Low, those dollars had been, in the days of the horse, not children's playthings but essentials in the ploughman's survival kit. The dollars were put by the ploughman between the cheeks of his posterior where they rolled gently to the rhythm of the horse and kept the sensitive skin free from rubbing. In really hot weather the horseman would have two sets of dollars. He would put one set in place and put the other on the dyke to cool. Then he would set off down the field. When he came back he changed his hot dollars for nice cool ones.

James Low told me about the feeing markets. He said that they were like livestock markets with men 'bought and sold like cattle', an expression I don't like but which must have been in some way correct because so many people used it. He told me of fighting in the boxing booths at those half-yearly feeing markets. The farm lads got a whole pound, a week's pay, if they could remain standing in the ring with one of the booth professionals for three rounds. James told me that he always entered and that he always got the pound. He had worked out a plan, which only a very focused, tough man could successfully have put into practice. He knew he couldn't knock the pro out so he didn't try. The crowd all wanted the challenger to win so the booth liked it when a farm servant doubled his week's pay. The mistake you had to avoid was hurting the pro in any way. James had seen so many lads sticking out their chests, weighing in with both hands and landing a lucky punch that hurt the pro, who promptly knocked them out. James did a certain amount of waving his arms about and grunting, to make it look like he was trying, but basically he just stood there and covered himself up to fend off as many blows as possible. Then it was off to the pub, where even a pound didn't last long.

At one market James made a bargain with a farmer and was ready to accept the half-crown of arles that sealed such bargains when the farmer said, 'Aye, well now. I'll need to speir about and get your character. So I'll meet you here in an hour.' When the hour was up the farmer said, 'Now that's grand, I've got your character and it's all good so I'll expect you home on Saturday night.'

'Aye well,' said Jimmy, 'but I've had an hour to get your character, and I'm nae comin.'

Jimmy was, unusually for Aberdeenshire farm servants, a socialist, but he was also just about the most reactionary person I ever discussed things with. He was in favour of corporal punishment and when it came to the birch, which he thought would be wonderful for keeping law and order, he would gladly have rolled up his enormous thick sleeves and laid it on himself. He would have had no hesitation if asked to tie the rope for any of the many classes of criminal he thought deserved to hang. He believed in unemployment benefit and social services for unfortunates, but the birch would have been good enough for 'buggers that winna work'. With views like that, maybe I shouldn't call James Low a socialist. But he was a fervent member of the Labour Party and insisted that all the men at Little Ardo join the Farm Servants' Union. His mind had been made up by a former employer, Mr Brebner, when he was fee-ed at Balquhindachy of Methlick about 1924. Those were the days when there was a clear class difference between the parties. The Conservatives or Unionists stood for the bosses and Labour stood for the workers, though in the case of the Scottish farm workers, most voted Conservative. At dinner-time on the day of Jimmy's first vote the boss, who was already fou, said to the young men, 'Now boys, you'll vote for your employer and you'll get bigger the wage.' Jimmy reckoned that if his employer wanted him to vote Tory his interest must lie with Labour. He never wavered.

By the time he was working for my family, James Low was a very loyal employee and would do anything the boss told him. One time, Maitland Mackie had bought a large consignment of Ayrshire steers from the dairies of Canada for fattening. They were as much as four years old and had nasty pointed horns, as Ayrshire cattle do. They had done quite a bit of damage to one another with those horns on the train across Canada, the boat across the Atlantic and then the train again to Arnage Station, some six miles away (Jimmy had collected them from Arnage himself and walked them home to Little Ardo). When Maitland came from North Ythsie to see the stots he was well enough pleased with his purchase but told his grieve to 'get those horns off'. They had no handling facilities, with races and restraining gates; each beast was lassooed and tied to a post, then James Low had to cut the horns off with a huge pair of shears called a 'breem-cutter', usually used for cutting stems of broom. I have done it with anaesthetic to a properly restrained yearling and I didn't like it one bit, but Jimmy had to do it to four-year-olds with great brutes of horns . . . without anaesthetic. When the man got the first horn off, the beast fainted, which made the second horn easier to get. Blood spouted everywhere. By the time the forty steers had been dishorned the midden at Little Ardo was awash with blood. Jimmy was terrified that he would land in Peterhead prison because even in the 1930s there were laws against cruelty to animals. He sent three men with a cart to dig up the road to Methlick and tell anyone who tried to come to Little Ardo that they were putting in a fancy drain. By the end of the second day the bleeding had stopped, none of the steers was any the worse and Jimmy had got the midden cleaned. He got the men to fill in their 'drain' and open the road again.

Another occasion when James showed his devotion to his employer was when Little Ardo had to be tested by the sanitary to get a licence to produce milk. John Allan once said that the public health people regarded milk not as food, but only as a source of infection. Certainly they were very particular about which farms got licences. I don't know much about those tests but James did tell me that one of them was that all the cow's water had to be piped. There were to be no waterins (fenced off bits of rivers or streams for the cattle to drink at). James showed something between pride and wonder when he told me that he had had to tell thirty-four lies to the sanitary man before Maitland Mackie could produce milk at Little Ardo. Again he feared Peterhead prison and dreaded the knock on the door, but luckily it never came.

I like to boast that Little Ardo has been farmed for six generations of my family. We started with William Yull and his son John in the nineteenth century followed briefly at the beginning of the twentieth by his grandson George Yull and then by Maitland Mackie, who had by this time married George's sister Mary, my granny. Then came Maitland Mackie's son-in-law John Allan and grandson (myself) and I have now been followed by Neil Purdie, who has married my daughter Sarah. They have two daughters so we can even say there have been seven generations now on the place since 1837.

Yet in my father's time, between 1945 and 1973, the husbandry of the place was almost entirely in the hands of the grieve. A wise man once said, and it has been repeated often, that the Aberdeenshire grieves were a race of potential prime ministers with far more important things to do. James Low was certainly one of those. He was an organiser and a planner. The binder was always ready the day before the corn, with its blade sharp and its cloths repaired. In difficult years, when fools would have rushed in and taken the crop half ready or wet, he would wait for the day, which always seemed to come eventually, and make the best job in the district. When the storm came in winter there was always plenty of corn in the loft, straw in the barn and neeps in the neep shed. In the summer when cattle broke out or even threatened, there was never any panic looking for the mell hammer or a few paling posts. James kept a cart loaded all summer with all that was needed to repair a broken gate, replace a strainer post, or mend a fence.

James Low was good, he knew he was good and he didn't mind who else knew it. As a young man he took particular pleasure in showing up his elders and betters, and later he much enjoyed giving his youngers and brashers the same treatment. I remember the time when, as an old man, he watched with delight as four lads from a neighbouring farm who had had the effrontery to rent the grass at Little Ardo, tried to get forty heifers out of the field at the bottom of the big brae. They were trying to get the beasts up to the farm for worming but they were hopelessly aggressive and in waving their arms and whooping at the ladies they just caused chaos. Time and again they got them near to the gate only to spook them and send them careering down again to the bottom of the field. After a couple of hours of this rodeo the four lads retired to the pub for a bite of lunch.

When they had gone, shutting the gate behind them, James re-opened the gate and called out, 'Come on then. Come awa lassies.' Then he stood back down the road a bit and lit his pipe. The curious heifers, intrigued by what might lie through this gate and, no doubt, the thought that they maybe weren't supposed to go through it after all, came tentatively out and started grazing the sweet unfertilised grass on the roadside. The grieve quietly shut the gate and walked them up the road. The men were astonished to find the beasts penned and waiting for them when they returned, to continue the fray, at one o'clock.

Jim Brown, who was fee-ed at Balquhindachy of Methlick when James Low was third horseman there in 1923, told me one of the few anecdotes I know where Low came off anything less than best. He was already up to the tricks of showing up his betters. In the words of Geordie Morris's song written about the neighbouring place of Mains of Cairnorrie he was, 'Raisin ragnails on the Foreman's heels. He fairly kept his roons.' The roons in this story were loads of neeps carted from the field to the neep shed. The second horseman was a good man and willing but was steady rather than a go-getter like James Low. Low was third and was always showing the second up. It was so bad that the grieve, the famous Jock Wilson, who deserves a chapter to himself, used to try to keep the two apart by sending them in turns to do this job and that. So this day Low had been at the plough and the second had been at the neeps. In the afternoon the roles would be reversed and Low would take over the cairtin.

'Foo mony straik had ye this mornin then, Thomson?' asked the third of the second.

'Five.'

'Huh,' said the third.

At half past four with an hour still to work, the grieve, who had heard that exchange, saw Low walking his horse and cart briskly through the close and heading back to the field. 'Foo mony straik's that Jimmy?'

'Five but there's plenty of time for anither een,' said Low.

'Na, na,' said the grieve. 'That'll dae for the day. Gwa and get the breem dog and you and me'll get an hour at the breem.' He was ordering the horseman to spend the rest of the day along the back road pulling broom out by the roots, grabbing the broom bush near the ground and, putting his foot on a wooden bit at the bottom of the 'breem dog', levering the plant out of the ground. It was one of the many toils on the farms which in those days had to be done in the most difficult way possible. So James Low was thwarted and, not only that, he had to spend the rest of the yoking on the dourest of jobs, suitable only for loons and orramen. He didn't often come second best. But of course, at that time, though he was already a potential prime minister, James was not yet himself a grieve.

By the time he was a grieve, and no doubt most of the time before that,

James had the gift of getting the job right. But it wasn't a free gift. Most of his craft was learned in ways that were very tough indeed. The boy was only thirteen years old when in 1914 all the able-bodied lads between the ages of eighteen and thirty left the farms of Aberdeenshire to fight the Kaiser. And the war meant a greater demand for food and a desperate shortage of people to produce it. That meant the young James Low, who had left school the day he was thirteen, had a single beast to caw and by the time he was fourteen had a pair of enormous Clydesdales to work. And remember, this was a man who never grew to above five foot seven inches, and at that time was nothing like full-grown, though he had tremendous shoulders and that Pictish combination of short legs and long bodies that made generations of Lowland Scots stronger than they looked.

Being thrown into the work of a man when he only had the body of a boy gave James an early start in thinking about how to do his work. Like the time he was sent with his single horse and the bone-davey to sow bone manure.

The manure had been set out in the field for him but he would have to lift the bags up by hand and tip them into the bone-davey, a height not far short of his shoulder. But the bags weighed two hundredweight, fully 100 kilos, and there was no way that the boy could lift them. However, necessity mothered an ingenuity beyond that expected of a thirteen-year-old. The bone-davey had spoke wheels and James was able to stick one corner of the bag into the spokes. He then called to his shelt to move slowly forward. As he held the bag on, the precious fertiliser was lifted up until the boy could tip it into the hopper. With the bravado of youth Jimmy didn't let anyone know how he had done it. He preferred them to think he had lifted the bags and was now ready for a man's work and a man's wage.

By the time he was fifteen, James had ploughed a whole winter. It was no wonder that by the time he was twenty-three he should be known as 'Auld Low', and six years later be grieve at Little Ardo, and ready for the job. It is no wonder either that I have always been in such awe of this man, for I only know of one thing that he got really wrong. When old Maitland Mackie came down in 1928 to the potato ground at North Ythsie where James was ploughing and asked if, after the term, he would start as grieve at Little Ardo, Low was of course delighted. Promotion had come more quickly than he had expected. He had known that he'd be 'spiered' (asked to stay on at the term) but just to stay on as second horseman. So here he was to be a grieve without first having been a foreman.

So the promotion was fine but the place was not to James Low's liking. As he walked up the Big Brae the place made no better an impression on the new grieve than it did on me when my mother and I arrived fifteen years later. It wasn't the lack of shelter for man or bird that disappointed James Low. It was the state of the steading, which he described as 'done'. 'A year'll dae me at this place,' he said to himself. In that last he was quite wrong. Forty-four years later, when he finally left Little Ardo, the old steading was still done, but it was still there.

Little Ardo's new grieve brought to the job all that was best and a bit of what was worst in the Aberdeenshire farm servant, in the days before they became farm workers or farm labourers and started to disappear. First and last was the passionate devotion to the job and the mission to do it right, do it better than the neighbours and, when mistakes were made, make sure no one saw them. A mistake beside the road was twenty times as bad as one in a shed. The great competition to be started and finished first with the very public jobs of hoeing turnips and harvesting were a universal passion. And James added his own sophistication to that over the years. By the '40s and '50s, when others were losing out by taking corn which was too green just to be first, Little Ardo was getting the reputation of starting last but finishing first. The moment any job was finished for the year Auld Low, as he was often called, would find an excuse to go to the shop or to the smiddy to see who he could see and greet with, 'Aye, aye Wardie. That's nae you still hairstin is it?' or, 'It's an awfu time o year tae be makin hay, Hillie. There canna be much in that stuff you're among eynoo.'

The best example of this passion for the land, and especially what other people were doing badly on it, was when the new power-driven binder came home. It was a wonder in its time: it could cut no fewer than six feet at once when full (though James never did quite fill it, reckoning shrewdly that the machine would last better if it wasn't pushed to its limit) and, being powered by the engine of the tractor rather than being dependent on its own forward motion, it could be run very slowly through the thickest crop and take on corn that had been flattened by rain or twisted by the wind that the old trailing balers hadn't had a hope of getting through. Despite all that power, it wasn't long before the new toy did stick in a tangle of flat crop which just wouldn't go into the cloths even with the tractor stationary and the binder going full-out. The tractorman put his foot on the clutch, which stopped both tractor and machine, and the grieve jumped off to clear the blockage. But the tractorman hadn't mastered the subtleties of the new 'live drive' and after nicking the tractor out of gear as usual, jumped off to give James a hand. That started the binder again and, to cut a long story short, the grieve got a broken leg.

The ambulance came eventually and James was put on a stretcher – whiles conscious and whiles unconscious. He was actually out when he was loaded, but no sooner had they closed the door than a tremendous racket got up in the ambulance. They rushed to open the door and see what had happened. They found the patient restored to his contemptuous best and struggling to rise. 'Dinna leave me lyin here. For God's sake, prop me up so I can see foo the hairst's gaun aa the wey tae the hospital.'

The farmer of Little Ardo got nothing but benefit from leaving the farming to James Low. I remember him producing the enormous sum of £3 and giving it to my father. It was for a load of muck, one of our little cartsful, bought by his eldest son, for his garden in the village. Had James Low Junior asked the farmer for a load of muck I'm sure he'd have got it for nothing. Every buyer was subjected to Low's dedication to the interests of Little Ardo. Neil Godsman, the farmer of Cairngall at Mintlaw and a tough guy himself, but with a sense of humour, was most taken by the grieve's salesmanship when he came to buy a cast cow at Little Ardo. He was only wanting it to rear a couple of calves so he wasn't that worried that only three of its teets were operational. 'By the time James had finished you'd have thought they had been breeding for three-titters since hine afore the war.'

John Allan was often astounded by his top hand's honesty and no more so than in respect of Mrs Low's hens. She had a well-run henhouse which we used to say produced enough money from the sale of surplus eggs to pay the Low household's food bills in the days after the war when eggs were so dear. Anyway, it didn't last and about 1956 James Low told his employer that Mrs Low was giving up the hens. 'They're jist nae payin.' John Allan had assumed that most of the expense of feeding Mrs Low's hens would have been borne by the farm, as was common on Aberdeenshire farms at the time, and was ashamed of having thought so.

The Aberdeenshire farm servants at the tail end of the war had a monstrous pride in their ability to work. James was proud, but he maybe wasn't as bad as Charlie Simmers, with whom I was cottared in 1959 at Banchory-Devenick. Though only just over five feet tall and weighing just one hundredweight, he was proud of the way he could carry a bag of wheat on his back up the stairs into the loft. Wheat bags in those days weighed a hundredweight and a half, one and a half times his own weight. And Charlie was so proud of his prowess in lifting such bags and carrying them up stairs that he got a job with the Angus Milling Company, where he would have to do so all day.

James wouldn't have done that, but he wouldn't have let anyone beat him at any job either. And his pride in his work took the form of contempt for those who couldn't do the job right or do it quickly. He saved special contempt for the workers on big estates. 'They just dae what they're tell't and they're aye up the erse o the laird or the factor.' He loved to tell me about the time they got a military man to be factor at the Haddo House Estate. The story illustrates his contempt for estate workers and the superiority of 'richt fairm men': The Haddo House men were draining a field to get rid of a patch which was always wet and often had standing water on it. On this occasion the wet bit amounted to a small loch. The plan was to put in a drain to take the water down to a stream some 100 yards away. The men started at the stream and were digging their way up hill to the water. When they reached the top they would carefully dig away the last bit of the drain and the water would start draining out. The level in the loch would drop and they'd dig out another bit and so on. If they were careful they could perform the whole operation with dry feet. But when the new factor saw what the men were doing he was scandalised. 'What the hell are you men doing? You're not taking the stream up to the water. You're taking the water down to the stream. Go up to the top, get in and dig the water down to the stream.' James Low used to swear that these men did as they were bidden. Up they went to the lochan and puddled their way down to the stream. 'Now, you would never get richt men tae dae a thing like that.'

Another feature of James Low was that he had no undue regard for the people society deemed his betters. He was not afraid to give my Uncle George Mackie, his boss's son, the mother and father of a hiding in the 1920s, although work was scarce and your employer had the power to make you destitute and homeless. Some employers did. Well, I was another farmer's son with whom James took what others might consider a liberty, though it was entirely to my advantage.

One of the crosses that James had to bear at Little Ardo was the drain that took the rainwater and the liquid waste from the midden, across the close and down to the wee burn in which it made its way down the brae, and into the Ythan and so to the sea. That drain would be quite illegal today but that is not the point. It had very little run on it and it was always blocking. And James seemed to spend half the winter trying to get it open and keep it that way. He had no drain rods. The tool he used was a double thickness of paling wire twisted together. He thrust this down the hole, twisting as he fed it in. When it came to the blockage he thrust it back and fore and eventually cleared the drain. On this occasion the drain was being particularly difficult and James was in a suitably foul mood. It did not improve his mood that he was being watched by the boss's son, and it drove him to distraction that the boy wouldn't stop fiddling with a milk tooth which had become slack. 'Leave that teeth alane, Charlie min.'

'Oh yes, Mr Low. Sorry Mr Low,' I said for I knew exactly what was bothering the great man. But somehow I soon forgot and I was fiddling with the tooth again.

'Stop that will ye.'

'Oh yes, Mr Low. Sorry Mr Low.'

Then we had 'Charlie min, ye bugger min. Stop that or I'll stop ye.'

But it was no use. The temptation to waggle a slack tooth was just too much . . . and the drain was as blocked as ever.

'Right,' roared the grieve. It was all over in a flash. He grabbed me and marched with me under his arm to the tractorshed. There he opened the toolbox of the old Fordson tractor with the bouncy seat and produced a pair of pliers. When I saw the pliers I understood and let out a scream of terror. That was all James Low needed. The mouth opened. In went the pliers. Out came the tooth. I can taste the mixture of grease and sand to this day. James still hadn't redd his drain but at least he had stopped the fiddling. It wasn't sore. The tooth was no more bother. It was a crime without a victim but I expect a grieve who did that today would face the sack and the police. Those were more practical days.

On the other side of the coin was the cruel sense of humour, so redolent of the bothy culture. It is one of the shames of our part of the world that our folklore is so lacking in the sort of taste which is now sometimes called political correctness. It is full of jokes about misfortune. The milkmaid who was so ugly she threatened to 'put the coos aa dry', the unrestrained abuse of the rival in love who had a 'nose for splittin hailsteens and a humpy back', and we are asked to find it funny that McGinnis should marry a 'cross-eyed pet'.

James Low had a bit of that sort of cruelty. Certainly the two lads who got James' present of a bottle of stout thought his sense of humour was pretty tough. The three of them had been doing some long and hot work in the Waterside Park on one of those rare summer days when the Ythanvale becomes a suntrap. The two lads were sent home with a load and the grieve went to Methlick on some errand. When he was in the general merchants he was bewitched by the sight of Guinness on sale and, in what started out as an uncharacteristically benevolent gesture, James bought three bottles.

Sadly the lads were a long time in returning and it was very hot. James couldn't wait for the others and drank his bottle. One of these little bottles of stout is not much use to a man of James Low's stature of a hot day so he drank the others as well. Well, they were his after all. No one could have faulted him if he had let it lie there, but as he told me, and I can't tell you whether it was with pride or remorse, he then filled two of the bottles with bursen oil from the tractor and replaced the caps.

When the lads returned they loaded his tractor, and as James was setting off for the steading he called out to them, 'Aye lads ye'll get twa bottles o' stout lyin ower there aside the dyke.' As he made off up the hill he heard one of the men remark with warmth, 'By God, James Low's a richt lad tae be fee-ed wi.' What they said after tasting the oil is not recorded nor, I suspect, could it be.

Then there was his treatment of the head of the McConnachie family of travelling people.

I was watching James Low trying to get a rusty towbar, which hadn't been touched for years, off the old hay turner. He wanted to take it to the smiddy to get a minor modification done and didn't want to be seen on the road with such a piece of agricultural history. The Little Ardo cows were all fed on silage, and so hay was considered very old-fashioned in Methlick in 1947. The grieve had tried elbow grease and he had tried oil. He had tried putting a metal pipe onto the shaft of the stilson to give him more leverage but he knew that he would just break the threads of the bolts if he used all that power. He was thinking he'd need to go for the blowlamp to give it a heat as the two-minute job stretched past the half-hour.

So the tink had chosen a bad time to call. Mr McConnachie was worthy of a civil welcome. He and his family were very important in the economics of Little Ardo farm. They bought heavy scrap and would take worthless light stuff away 'out of yer road'. They also dealt in rags. But the main role of the travellers was in maintenance. The seed potatoes spent the winter in tattie boxes. These were not the one tonne monsters used for handling spuds on twenty-first-century farms. They were more like trays than boxes. They were about one foot by two feet and they had a handle along the top. In April the seed potatoes were taken out to the field to be planted by an army of school bairns. Large numbers of these boxes were damaged each year and tinks made a first-rate job of mending them. They were even better when it came to mending the tattie skulls. These were wire baskets into which the kids on their tattie holidays gathered the potatoes for loading into the carts in October. The skulls were always in a dire state by the end of harvest. The children were always sitting on them and leaving them where they would be runover by the tractors. How the McConnachies did it I don't know, except that they had a frame a bit like a shoemaker's last, over which they fixed the battered skull to get it into shape, and from which it emerged as good as new.

But it was not a good time to call. 'Ah! Hallo Mr Low, it's good to see you. Been terrible weather and don't I know it.'

'Humph.'

'So would you have any owld scrap Mr Low, I could give you a good price today,' said Mr McConnachie in that curious accent which has a bit of the south-west and lot of Irish in it.

'We've plenty a scrap but we're still usin it,' said James, giving the hay turner a petulant kick.

'Oh yes, Mr Low. Quite so. Hard times it is, indeed. I quite understand.' And then after a slight pause Mr McConnachie tried again. 'Mr Low, I've got the wife here with me now, Mr Low. That's her in de van. Would you mind if she were to go to de door to see if Mrs Low would have any owld cloths she'd be throwing out that would do for de bairns?'

'I've bairns of my ain,' said James without a smile.

'Oh yes. That you have, Mr Low, quite so. Hard times indeed yes it is. Mr Low, ye wouldn't have a pair of owld boots would ye, Mr Low?'

James Low took a good look at McConnachie's boots. They were not a pretty sight. They were both down-at-heel, except that the right one had no heel at all and the left one would have been flippity-flap at the front if the sole had not been tied on with a piece of binder twine. 'Good God man,' he said, 'is that pair nae auld enough for ye?'

I am a bit ashamed of that story, although my only part in it was as an audience. I have told it often as an example of how tough people were in those days, and how cruel. It's a good example but I am not sure that I haven't told it so often because people in the North-east do still find it funny.

17
Mrs Low's Kitchen

It was not a farm kitchen in the old sense. It wasn't the living/dining-room/kitchen which is central to the bothy culture of the North-east. Not like the farm kitchen at my grandmother's home at North Ythsie and which now was in the Walkers' part of the house, where Mrs Walker had cooked for and fed her family and the single men and where, of a winter evening, the accordion and the trump could be brought out. In Mrs Low's kitchen the only things that went on were cooking, baking and washing. It was known as the back kitchen. The main kitchen was next door. There the grieve sat at the left side of the fire smoking his pipeful of Bogey roll and spitting without spilling a drop into the fire. Despite James's accuracy she thought this an uncouth way of doing and provided him with a spittoon lined with sawdust. It sat on the floor at the wall side of James's chair where the minister might not see it on his annual visit.

But it was the back kitchen which interested the young farmer. It is a curious little afterthought added on to the end of the Victorian wing of the old farmhouse, granite-built, with a proper pitched roof and slates. But it is an altogether humbler-looking structure than the rest of the house. It was as though they had taken a wee but and ben, cut it in half and stuck one half on to the farmhouse.

The Lows ate earlier than we Allans, so whenever Mrs Low had food on the go I was hungry. And she was a wonderful cook of all the standard meals on the farms in those days. But, before I tell you of all the great diets she

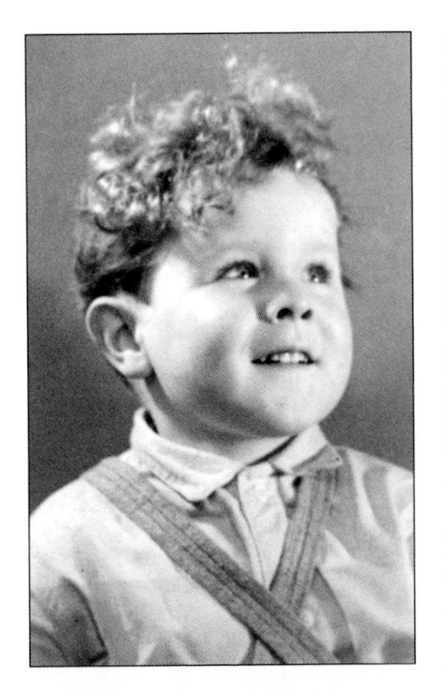

ABOVE LEFT. The author, Blairlogie, 1942. (All photographs © the author)

ABOVE RIGHT. The author with grandfather Maitland Mackie and the Riley Nine, North Ythsie, c.1943.

LEFT. Mary Mackie, the author's grandmother, with her first three children – Jean, the author's mother, John and Mike, c.1923.

ABOVE. Maitland and Mary with all their children (left to right) – John, Mary(jr) George, Catherine, Mike and the author's mother, Jean, c.1926.

LEFT. Aunt Catherine in 1952.

Mackie's Milk tug-o-war team, winners at Turra Show, 1936. James Low is second from the left, kneeling. Maitland Mackie (peeping) and Mike Mackie are standing on the left.

Little Ardo from the bottom of the Big Brae, c.1953.

Maitland and Mary's wedding party in the close at Little Ardo in 1906. The author's great-grandparents, both by this time widowed, flank the happy couple. John Mackie, father of the groom, sits next to the bride and Mrs John Yull sits next to the groom.

Mackie's Milk tug-o-war team, winners at Turra Show, 1936. James Low is second from the left, kneeling. Maitland Mackie (peeping) and Mike Mackie are standing on the left.

Little Ardo from the bottom of the Big Brae, c.1953.

Maitland and Mary's wedding party in the close at Little Ardo in 1906. The author's great-grandparents, both by this time widowed, flank the happy couple. John Mackie, father of the groom, sits next to the bride and Mrs John Yull sits next to the groom.

RIGHT. The author's great-grandfather, John Yull, c.1879.

BELOW. John Yull with Clatterin Jean and her foal.

LEFT. The author's grandmother, Mary Yull, well mounted on Donald, c.1893.

BELOW. The author's father, John Allan, drives a gig at Bodachra, 1912.

Alexander and Fanny Allan, 'the Auld Folks on the Wa'' – the author's great-grandparents who brought up his father, c. 1905.

The Allans of Bodachra, a composite photograph taken between 1870 and 1913. Chae and his elder daughter, Charles and Alexander. The author's eponymous great-uncle, cousin once removed, great-great-grandfather and great-grandfather.

The author's grandmother, Fanny Allan, with Fred Willows (c.1940), whom she met and married in Canada.

The Heroes of Andover. John Allan (centre) and his two army pals who taught him to drink off the piano and throw Yanks to the ducks.

Village hall election meeting, 1945. John Allan is far left and Bob Boothby is far right – just as it should be.

James Low with the second pair at North Ythsie, 1928.

James and Isabella Low with (left to right) Joe, Belle, Jimmy, Dod and Albert, 1945.

John Allan admires his elevator and chaff blower, 1948

Fine new concrete close (needing a sweep already), 1948.

Albert Low with the author in front of the Little Ardo piggery, 1945.

RIGHT. Scotland's first First Minister with the author in the garden of St Nicholas School, 1949.

BELOW. James and Mary Kelman in Little Ardo's byre, 1948.

Jake Clubb gets the milk away, 1947.

The Hairst Squad 1954. John Allan, the author, Craigie Taylor (sitting), Willie Adie, Bill Taylor, Mary Taylor, Jimmy Taylor (seated), Sarge Mackinnon and James Low.

RIGHT. Susan Rennie (née Allan), John Allan's protector, with her youngest son, Benji.

BELOW. John Allan, united at last with the rest of his mother's family – half sister Fanny, half brother Gordon Willows and two little nieces – in Canada, 1951

used to put up, I must use her mince and tatties to illustrate the difference between Mrs Low's cooking and the cooking for the farmer's table through in the main house. My mother had had a training in cookery, knew all about vegetables needing to be al dente, and meat to be served rare, not to mention with what wine. Mrs Low, on the other hand, had been taught by doing it at home and being fee-ed to work in farm kitchens from the age of fourteen. It so happens that I wrote down Mrs Low's typical day when she was fee-ed at Balquhindachy of Methlick in 1918, in the days when she was still Isabella Rennie.

Isabella's day started started at five-thirty when she had to milk three cows or six when it was the other maid's day off and she was left as toonkeeper. She got her own breakfast at the back of six along with the single men and the other maid and the two of them washed up. Then she had to make breakfast at eight for the 'ben the hoose folk' in her own ben-the-hoose kitchen, 'serve table' and wash up. Then she had to make the beds, 'teem the bedpans', hoover the carpets, dusting and sweeping till it was time to prepare and wait table for dinner at one o'clock. She then had to wash up again before getting on with the afternoon's work of washing, ironing and cleaning the silver. She had to produce a cup of tea (she called it a fly) at three o'clock and high tea at five, grab a cup or a bite for herself and wash up again. At six o'clock she had the cows to milk again, put hot water bottles (pigs) in the ben-the-hoose folk's beds, and make down the beds. She finished at eight o'clock and bed time was at ten unless there was a party on and more serving and washing to do. Isabella had a sixteen-and-a-half-hour day thirteen days in the fortnight, twenty-six fortnights a year. Her wages for that were £8 per half-year, one shilling paid as arles to seal the contract and £7 19s at the term. And Isabella had to ask Mrs Brebner, the farmer's wife, for permission if she wanted to go to a dance in Methlick, just three miles away.

When Isabella became Mrs James Low she lost her Christian name almost completely. She became Mrs Low. It was the same with all the wives on the farms of Aberdeenshire before the war and for a long time thereafter. I fear that it was a symbol of male dominance, and James Low was definitely dominant in their family. But not in the kitchen. In fact, Mrs Low had the run of the house and for that she and the children got all the domestic work to do. He was quite an old man before I saw the grieve do a hand's turn in the house and that was only to pour the tea.

My mother's training at the fancy cookery school might have been handy in the restaurants of Mayfair but I have no doubt that Mrs Low's training at the cookery school at Balquhindachy was the better for the North-east of Scotland just after Hitler's war. My mother used to go to the butchers and order a pound of Mr Presly's best steak. Having made sure it was trimmed of most of the fat she would then tell him to put that through the mincer. This was Jean Allan's mince. She produced it some time between one and two o'clock along with some onion chunks and warmed up wheels of carrot. This, applied to mashed potato which often didn't have many uncooked lumps in it, was one of my favourites among my mother's meals for the three of us.

By contrast Mrs Low would send one of her sons to the butcher for three quarters of a pound of his ordinary mince, which everyone knew comprised basically everything that couldn't be sold; the trimmings, the fat, and anything that hadn't sold the day before and the day before that. To this she added a lot of water, some gravy browning, a suppy flour for thickening, a heap of chopped carrots and onions and plenty of salt and then boiled it slowly for an hour. Sometimes she would add an oatmeal pudding, sometimes called a mealy Jimmy, to be shared among the diners. So, you got a great heap of wonderfully flourie-dry mashed potatoes and your portion of mealie pudding. Then you helped yourself to a couple of spoonfuls of mince from a big bowl on the table. With all the oatcakes you wanted, that was a feast for her family of seven and there was even enough sometimes for the farmer's son who was always hanging about.

She made wonderful potato soup which we all called 'soup tatties' and stoved potatoes (or 'stovies'), which used up any left-over beef and roast fat. Then there was hairy tatties. For this she took a very small amount of dried cod, not as much as half a pound, and boiled it. It was then beaten in amongst the heap of mashed potatoes. The cod was completely dissected into fibres of muscle, which stuck out of the potato, giving it the appearance of being covered in white hair, and eaten with mustard that made it nippy. I think they were my favourites – until I remember the soup tatties. The grieve's wife could even make a delicious meal without any meat at all. Many's the time the family sat down to just potatoes with a little butter or margarine.

Other dishes were not so popular with the young farmer. I didn't like brose (porridge made with boiling water and no cooking). Cabbage brose was a favourite with some and there the water was what the cabbage had been cooked in, and kale brose was made with the bree from the boiled kale. There was also brose made not with oatmeal but with peasmeal. I only had them once but 'pizzers' was not a diet for kings, or even for a farmer's son. Of course, the trouble with brose of all kinds was that the diner had to make his own. I am sure if Mrs Low had made them they would have been delicious.

In truth, I didn't often dine at Mrs Low's table but what I did get often was some of her baking. Every week and perhaps more than once she did a baking, a huge baking, for every fly cup and every evening meal relied on her baking and every single meal needed an accompaniment of oatcakes. There are two funny things she taught me about oatcakes. They are plural, for a start, as are porridge. Granny Low would say 'come on in to your porridge. They're ready.' And the second thing was that oatcakes were called breid. Bread was called loaf.

The baking was done in her little back kitchen or in the oven of the range. Mostly she baked on the open fire. Oatcakes were made on the girdle, which was suspended above the flame on the sway which swung off the fire to be loaded and unloaded and back onto the fire for cooking. The oats and water were mixed and kneaded by hand and then rolled out to the right thickness, with plenty of oatmeal to keep the surface dry and make it workable. Then the dough was put on the girdle and trimmed to cover it exactly. The girdle was put back on the sway and swung round so that it was on the fire and the breid cooking. When it was cooked Mrs Low cut it into quarters, sometimes called corters, which strangely were not quarters but eighths. They were then taken off the girdle and set up round both sides of the open hearth to 'fire'. That was to dry further and, depending on taste, they could even be toasted a bit. As each new batch was cut into corters they were set up in front of the others round the fire. It would be nothing to have them sitting six-deep all round the great open fire.

When Mrs Low was baking she filled the close with the most delicious savoury smell. When all's said and done they were only thin toasted brose, but they were delicious. And the farmer's son was never denied a quarter of breid. As an old woman who had left the open fire and sway at Little Ardo, Mrs Low continued to make oatcakes and she always had one for me, but she admitted that cooked on an electric hot plate and with no hearth in which to fire them, they were good but just 'nae the same'.

Mrs Low's catering was based on that hearthful of breid and on potatoes. One of my abiding memories of her kitchen was her two youngest sons, my pals Albert and Joe, on the floor peeling potatoes. They had quite filled one pail but their mother insisted she would need 'a fyow mair'.

Mrs Low's meals were finished with a milk pudding of some sort, semolina most often, served with a small amount of rhubarb or her own jam, raspberries or blueberries which we loons picked from the roadsides or the woodie, or rhubarb from the garden. Arthur Glennie, who farmed Hillhead, our neighbours to the north, between the '60s and the '90s, used to say that there are two things you get for nothing on a farm: rhubarb and kittlins.

Both were worth something. The cats kept the rats down, but the rhubarb was a real asset. Mrs Low's rhubarb jam, which was made with just enough ginger, was wonderful on a pan loaf or for topping off a rice pudding. It was almost as good as her girdle scones with golden syrup. Aye but steady on! For all the magic she conjured up there, Mrs Low's back kitchen was a dingy place, though I didn't notice at the time. It was lit by the door, which was open in all but the dirtiest weather, and a very small sash and casement window to the west. The walls were cream and all the woodwork was dark brown, either stain or paint. Certainly the meal girnal was brown-painted. That was where the meal was stored. It stood three feet high and was just a box which was filled to the brim perhaps twice a year, with perhaps a couple of hundredweights of oatmeal. In the girnal it would be safe from the mice which got into everything else.

But the great virtue of the girnal was that it kept the meal fresh and that was very important. Oatcakes were far better when the oats were fresh from the mill. Everyone had a couple of extra quarters when there was a fresh batch of meal in. It was important to pack the girnal tight with oats to keep them fresh. You had to tramp them down hard to squeeze all the air out so that the aerobic bacteria which turn them sour were starved of oxygen. I never did it and I never saw it done but I was told that the ideal way to pack a girnal was to get someone with their bare feet to tramp it while others filled it.

By the little window in the west wall were twin sinks, those great deep and wide things that served for the cooking, washing up and for the clothes washing. Those sinks and a washboard with plenty of elbowgrease were all that Mrs Low had at first for the washing for her family of five children plus the grieve and herself. Then she got a jiffy washing machine. That was a single tub with a paddle operated by a handle. Albert and Joe were lucky indeed that they were often required to caw that handle for ages on a Monday. Then there was a little mangle bolted on beside the sinks, providing more work for boys. Latterly there were hooks in the ceiling for a clothesline should Monday not be fair or in winter if there was no force to dry it. Beside the two great sinks there was a small sink in which Jimmy washed his head; he didn't just wash his face like my father did – after his work he really did wash his whole head. Completing the west wall was a hand pump, with which water was drawn from the old well in the close and pumped up into the loft from where it fed the kitchen and the famous water closet. And Mrs Low's back kitchen even had hot running water. That came from a back-boiler in the metal range in the old farm kitchen through in the main house and was stored in a cylinder in the west wall above the sinks. There was a big dresser on the south wall for kitchen utensils, most of the crockery being kept through the house, where the eating was done, and shelves and hooks for the pans on the south wall. There was the table at which Mrs Low did her baking and which she and her helpers ate when the kitchen was full of up to eighteen when there was a hash on and the steam mill came to thrash the Little Ardo grain. The north wall was dominated by the huge open fire with its sway and the east wall by the meal girnal. During the war and before it, that was a huge wooden box made of rough timber with the bark still on it. It took four bows (bolls) or five hundredweights at a time of the oatmeal that was the raw material of the brose, and of the breid. As part of John Allan's first improvements after the war a pantry with a window was made in the north-east neuk of Mrs Low's back kitchen and that meant the old meal girnal was replaced by a modest affair made with finished wood painted dark brown which held no more than two bows of meal.

And that is all that I can remember of Mrs Low's kitchen. Except the fly-papers. It is one of the things that differs most in the farm kitchens now compared with then. In the 1940s farm kitchens swarmed with houseflies. I don't know why there were so many flies in those days but I am very aware of how true it was and I think the reason why there are so many fewer birds these days is something to do with the fact that there are so many fewer flies for them to eat.

There was no refrigerator until much later in Mrs Low's kitchen, which meant that milk went off in a long day, while bread went off in three days despite being kept in a bread bin. Any uncooked meat went off in a couple of days. It also meant that the flies could smell the thing which attracts them best, raw meat, and they swarmed in through the ever-open door. There were always perhaps twenty flies circling endlessly round the single 100-watt lamp which hung in the centre of Mrs Low's kitchen. Well it wasn't quite endlessly because also suspended from that lamp was a flypaper. This was the stickiest thing in the world. About an inch and a half wide by perhaps eighteen inches long, it hung down and the flies stuck to it. It was not an edifying sight. The gluey paper was a sort of browny-yellow and it gradually turned to black as more and more flies became trapped. If you looked closely you could see little movements, which meant the flies were alive though their future was bleak. When the fly-paper was so covered in flies that some of the insects could land on the other bodies and take off again without becoming stuck, it was thrown out and another fly-paper unwound.

These were ugly things and it never cleared out the flies, though there would certainly have been several thousand flies caught by each paper. But my mother used various chemical means to achieve the same result. She shut the doors, covered the food and then skooshed DDT at them. That killed them all and we were choked but fly-free for a few minutes. I have no doubt Mrs Low's was the healthier option – however grotesque it may have been.

18
Me and My Loons

Of great importance to me were my loons – my pals. Here the team was six. There were the grieve's sons, Joe and Albert, and the cattlemen's three, Jimmy, Billy and Bertie. You couldn't count Johnny Kelman because he was too little, or Lil Kelman, because she had already left home. Only the two youngest Lows counted because Jim, Belle and Dod were too big and were already working. And you could hardly count the Taylors either. They lived at what had been the Smiddy Croft more than a mile away, and it was an important mile because it meant they went to Cairnorrie school and their nearest general merchants was Cheyne's at Cairnorrie rather than Grant's or French's in the village. There were other loons from time to time. Jake Clubb, a foreman, had two sons, though they were a bit too young. But Alex Rattray, second tractorman for a while, had a very earnest son of a better age called Robbie, and Peter Glennie, another tractorman who didn't stay long, had a son, Ian, who was a very able footballer and who was to earn undying fame in the village by kicking a visiting football fan, who had asked for it, on her backside.

What did I and my pals do to fill our days in the ten years after Hitler's war? Were we bored? Did we require an army of social workers to entertain us? How did we manage without our mothers driving us around to organised leisure activities? After 60 years, not everything can be remembered completely, but it is quite clear, and the four loons I still see back this up: boredom outside of the classroom did not exist for us. We were busy all the time, playing games and playing at being adults. All the parents had to do was to provide clothes and meals, and kick us outside.

We played all the usual games of Takye (tig) in variations including one where if you were tigged you had to stand with your arms outstretched until released by another player, Ye Canna Cross the River, Bats and Basies (rounders), Bobbies and Burglars, Hide and Seek, Kick the Cannie and Heist the Green Flag, and of course all those singing and skipping games. There were seasonal orgies of conkers, stibble ruck-building, sledging and bools (marbles). Then there were our own make-ups like a mixture of hide and seek and takye where teams of three had to find one another and then tig them. It never had a name, but we played that one for hours in the winter nights. There were opportunistic games like diving off the loft steps onto the stored straw and infuriating Kelman, the dairy cattleman, by messing up his tidy barn. Quite the best of these brings me out in a sweat to this day when I think of it, for I am quite claustrophobic. We made chutes from the top of the loose straw in the barn down to the ground. This involved one person standing on the floor of the barn and pulling out handfuls of straw and thus burrowing upwards. That job was a bit scary when you got in a bit but the real terror came to the one who started at the top and burrowed downwards. Again it was OK for a start. But soon the burrower was deep down in the straw passing handfuls of straw back between his legs and sinking ever deeper.

Before long the one who had started from the top was past the point of no return. Then the only hope was to keep on digging. Eventually the one who was digging upwards would touch fingers with the other who was by this time short of air and very short of space. The two worked away at it until the one below could get a firm hold of the other and then pull. Sometimes the pulling went on for some time but eventually, just like a stiff calving, out he would come.

I can hardly believe it looking back, but what happened next was a race to see who could be the next to dive down the hole, get stuck, try not to panic and be pulled out. It got easier and easier until it really was just like those tube-like chutes you see at some swimming pools nowadays. The only difference was that at the bottom of our chute we had a concrete floor rather than water.

Then we had games playing at imitating adult life. We put on concerts.

We ran a football league among ourselves. OK, there were only two teams of three, but the latest league positions were posted on the stable door and the trophy awarded at the end of the season was the Ardo Shield, a very fancy piece of polished mahogony. We played at being our mothers with lames (broken bits of crockery) for plates, food tins for cups, sand for sugar, souricks (sorrel) for cabbage and ripe docken seeds for mince. We made our own gardens where the other loons grew beetroot which they took proudly to their mothers for bottling. I favoured radishes. My loons thought that a very funny choice as they didn't eat salad. We made things in the farm shop-pie and we eventually made our own workshop in the abandoned henhouse in the wood. We had a very unsatisfactory selection of tools with which we made hutches for our livestock (rabbits and guinea pigs, which were for sale and were sometimes bought). We made furniture for our housies, several of which were up trees. There were no quines, which would have made 'Mannies and Wifies' not to mention 'Doctors' a lot more interesting, but one thing is sure, we were never bored.

Like the kids today we ate all the time. But we didn't need the shop. We ate neeps, bashed on the dyke to let us into the meat. Potatoes cooked on the braziers in the tattie shed. We ate rasps, brambles, blaeberries, arnuts, souricks, birdie's pizz and ripening grain. In October, if you knew where to look, you could get hazelnuts on the Braes of Gight. Another seasonal favourite was grass. Of course, not even we chomped up at mouthfuls of grass like a cow but, just as the cocksfoot is flowering in the month of June, if you take a hold of the flowering head and pull it up gently it will break above the last node and there will be an inch or more of delicious soft stem. It is difficult to get a whole mouthful, let alone a bellyful, but as we played football in the fields, or went on our expeditions, for one month in the year, we did eat grass.

As well as our football league, we even had an International. Old Maitland Mackie rounded up ten of the loons, including my special friends from my days at North Ythsie, Brian and Tondy Crombie, and the Goodalls, ringers from South Ythsie across the road, and brought them the eight miles to Methlick for a grand challenge match. The referee was my father, John Allan, the goals were fence posts and there were no lines nor cross bars. The home team could only field eight players but they included Joe Low who in ten years would be playing for the British Army of the Rhine, Ian

Glennie who, during his National Service, played in the English Third Division, North, and Charlie Allan who eight years later would be finding out that he wasn't as good as he thought, training with the Gentleman George Hamilton, Bobby Wishart, Harry Yorston and Paddy Bucklay at Pittodrie. Little Ardo won 8–3 and John Allan learned an important lesson. He had laid in refreshments: sixteen big bottles of lemonade of various colours, four dozen baps and two dozen scones. When those were consumed we ate the entire contents of the farmer's bread and cake tins. 'If you're bothered with stale food, have a football match.'

We put on concerts for ourselves in the stable loft. And we put on plays which were made up as they went on. We ran our own Highland Games.

We were hunters as well as gatherers. With wire snares we caught rabbits and hares and gave them to our mothers or sold them to the butcher. I once got seven shillings for a hare. We trapped pigeons in the steadings by shutting the doors and even stuffing the broken windows with straw. Then we chased them until they dropped. We plucked and gutted them and presented them to our mothers. Then there was fishing for trout in the wee burns. During the war there was the search for lapwings' eggs. They were good to eat but were left alone after the war as the peesies were getting scarce and the shortages of proteins to eat gradually eased. We all collected birds' eggs.

We all had bikes and there were days away. To the Middle Lake to swim. To Gight Castle to play at Robin Hood among the ruins. And to Turriff, fifteen miles away, to play a football match against the local loons. That was all organised by ourselves . . . and not an adult involved on either side.

It was boring at school but as soon as you got out there just weren't enough hours in a day. But how were we controlled? Why did we not burn everything down if it wouldn't break and we couldn't steal it? The Health and Safety Executive would have stopped half of what we got up to about the farm. All right, Joe Low did lose a finger but we all believe that he would still have the finger if he hadn't been warned of what happens to loons who put their finger in the cake-breaker. And I had to pull Tibber McBain out of the Middle Lake at Haddo House because he had jumped in confident he would manage to swim once he was in. He was right but he couldn't turn and was heading out across the lake; I was standing on a flimsy little jetty that stuck out into the lake and Tibber came swimming up to me and then past, heading out to sea. 'Look at at. Look at at. I can sweem. I can sweem,' he shouted excitedly. But as the cross-lake swimmer drew level with me his excitement changed to something more like panic. 'Foo dive ye turn? I canna turn. I canna turn.' At first I thought he was joking but when he started to sink and thrash about wildly I realised something was up and jumped in to the rescue. I had not only been taught to swim in the Uptown Baths in Aberdeen, but I had done a junior life-saving course: I knew how to approach the drownee. You are supposed to bash them as hard as you can in order to daze them in case in their panic they pull you down under the waves. Then you have to turn the semi-conscious victim round and, doing an elegant sidestroke, set off for shore pulling him by the head and making sure he is above the surface. Luckily for Tibber, all that stuff was forgotten. I just grabbed him and threw him towards the shore. Then I grabbed him again and threw him a bit farther. From there he was able to make his escape. Being left to ourselves we quickly learned what was sore. We also learned what our parents would accept if they found out. We were kept within reason by a well-educated fear of pain and not by violence – but perhaps by the threat of violence.

I can only remember one case of violence being used to reinforce authority. This lad was caught by the grieve in the act of breaking his thirty-seventh window pane at the back of the tattie shed. Now in the woodie beside the house at Little Ardo there is a rather handsome stone-and-slate-built granite building seven feet square, and with no windows. It had been built as a cold store for milk and other household perishables. For many years it had been used as a coal shed by the grieve and, with its lack of windows, with the heavy door shut it was absolutely black inside. The grieve shut the boy in the coal shed until his cries of terror died down to a whimper. Then he let the boy out saying, 'Now Colin, if I ever catch ye at the fairm again, I'll pit ye in there and I'll niver let you oot.' He would certainly be jailed if he did anything like that today – but it worked. We loons were all impressed.

We bairns all liked nothing so well as being able to do a man's work even if we didn't get paid. We really liked sweeling the byre. This was part of the dairyman's twice-daily routine. Using four-gallon pails, filled at the troughs at either end of the byre, he had to splash the floor with water to clean it. We loons with one splash of a full pail could make several yards of the byre floor spotless. We soon became experts at herding cattle, at knowing just what angle you had to make between the next herder and the cattle to show them where you wanted them to go, and even the youngest could be positioned to substitute for a gate. The tatties were a chance to pick up wealth that could only be understood by the bigger ones who had been paid the previous year. As soon as we were fit to do so without getting in the way we were eager to help at anything in the field, the byre or the garden.

It could be dangerous, as Billy Kelman found out. He liked and was entrusted with the job of tramping the silage pits. That is running the tractor back and fore across the pit to consolidate the fodder. When you are grown-up it is the most boring job there is, but Billy was the envy of us younger loons as he drove across and back. We all aspired to his important position. With a queue of willing reserves there was no need for wages. Billy was getting on fine this day. The silage was well up, perhaps seven feet high, when Albert and I brought our envy round the corner. Billy was already enjoying himself but our presence was exciting. He pulled the throttle out a wee bit, with many glances in our direction, and in reversing went nearer than usual to the edge. He started slipping. The old Grey Lady was clearly going to capsize, and Billy had no cab and no roll bar. He was in the greatest of danger of being squashed. However, panic produced a great surge of athleticism from the eleven-year-old and he was able to scramble and leap over the tractor as she fell. Only his pride was hurt but the damage there was severe. On the farm, even boys must never let their rivals witness their disasters.

I have not said enough about the ploys of the youth of Little Ardo. Hardly anything about the rabbits, guinea pigs, white mice, not to mention the trout we caught and kept in a pond centred on John Yull's well in the close. But our pet-keeping wasn't just a matter of having cuddly little creatures to fraise with or to show off. For all that my father was a bit dismissive at first he began to see that we were on to something with the guinea pigs. We bought three for five bob in April 1949 and we were blessed with an increase first of three and then of four. And by June Albert and I were writing to the head of the Rowett Institute for Research, which studies nutrition, to see if he needed to add to his herd. We got the following very nice letter back. It was addressed to Messrs C.M. Allan and A. Low, Breeders of Cavies:

Dear Sirs,
Thank you for your letter informing us that you have some cavies for sale. We breed adequate numbers to meet our needs but it is interesting to note that an alternative supply might be available should an epidemic or violence wipe out our stock. Dr J. Smith of the City Fever Hospital might be glad to know that you have some stock for sale and I suggest you write to him.
When our new animal house is ready you might like to pay us a visit,
Yours faithfully
D.P. Cuthbertson

Professor Cuthbertson was quite right. And so it was that ten days later Albert and I were at the City Hospital with nine guinea pigs in a cardboard box. We were expecting half a crown and hoping for three shillings. We were well taught in negotiation to go to any length to avoid making the first bid and how right that advice had been. Dr Smith spoke first and offered four shillings apiece. We were overpowered. We'd fairly shown the farmer how to make money and how to turn over our cash. Better than that, when we got home to Little Ardo having sold my four pigs and Albert's five at a handsome price, I discovered that my junior sow had delivered herself of a litter of five, another pound's worth, and all were doing well. Then I could march up to Nat Milne's grocers' van when next he came on his weekly visit to Mrs Low and plank down a whole penny. For that I got a whole mealy Jimmy and a farthing change.

We did feel a little remorse at our guineas going for medical research but business is business. As the farmer wrote in his diary: 'The Little Ardo guinea pig breeders wouldn't have changed places with Culrossie.' (Leading cattle breeders of the day.)

One day in 1946 we loons felt we were all millionaires. A lorry driver who had been delivering a load, I can't remember of what (to us loons it didn't matter) suddenly threw down from his lorry a large number of cans of corned beef. In the days of food rationing – remember the meat ration fell as low as two ounces a week – this was fabulous wealth. The Lows and the Kelmans got two tins and the rest of us got a tin apiece and we bore them off in triumph to our mothers. There is no way it was legal but even my parents, who would never take anything off the black market, didn't resist. Even if it was black, it was free. Somehow it wasn't so corrupt to take black stuff that was free. It is the only time I have seen things literally falling off the back of a lorry, though I'm told it is common.

Another of our excitements came when the butcher from Maud decided to include Little Ardo in his round. Maud is eight miles away so he was putting the boat well out. The young Lows had heard their parents discussing this development and were quite alarmed. The local pronunciation of Maud is 'mad', so we were on tenterhooks all morning the day the mad butcher was due. We climbed up to the tops of rucks in the cornyard, partly to be safe and partly in the hope of seeing him. Sadly we couldn't quite see him, Mrs Low survived and, so far as I know, he never came back.

Even the Maud butcher included, there was nothing so exciting in our lives as the football matches. Methlick play in the Buchan League against all the big teams like New Deer, Ardallie and New Pitsligo. Each week throughout the summer there were, as there still are today, two evening games, one at Methlick and one away. Evenings were the only times possible then, for the farm lads didn't get off on Saturdays and the Kirk made sure nothing happened in full view of the road on a Sunday. There was always a bus that took the players and spectators to the away matches but we loons didn't often get to away matches. There weren't enough seats on the buses. You had to pay a shilling for your fare, or even one and six if it was far away like Rosehearty. We didn't like Rosehearty. There were at least two reasons for that. They were very good and were rumoured to sneak minor Highland League players into their team. While we had to get all our players from Methlick Parish as stated in the Buchan League's rules, Rosehearty somehow managed to get their players from Fraserburgh. The second reason was racist. They were fishers and as such were a different breed. They had nothing in common with us farming folk and weren't nearly as good.

It wasn't just the loons who thought the return of the Buchan League a great advance in civilisation. The crowds in the summer evenings at Methlick in the 1940s would have compared favourably with the Highland League in the 2000s. There was no turnstile to count them, the hall keeper just went round with a collecting tin. But all the village loons were there and a few of the quines and with the all-male adult audience about four-deep most of the way along one side of the pitch and a busload of visitors usually one-deep halfway along the other side, 300 or 400 would have been an average crowd on a fine summer's night.

We loons were really very well behaved within our own community but when it came to fishers coming to Methlick to play fitba and beat us, we could be pretty offensive. Excitement was high. Methlick played better than usual and the visiting spectators were visibly worried. There were only a few minutes to go and the teams level at four apiece, when Rosehearty scored to a roar from their busload of supporters and a good deal of swearing from the rest of the crowd.

But then to all our astonishments Methlick equalised. It was five each. We loons were already about cheered out but even the old men who were only there to criticise and tell anybody who would listen that the games were never so good after the war, could scarce forbear to cheer. It was five all and that would show the fishers who thought they were so good.

It is just as well the bobby wasn't about for one of the visiting ladies took down her knickers, lifted her skirt and bared her considerable posterior to show the good people of Methlick what she thought of their football team. She was just a few yards from us loons so we got a good look. Most of us were far too busy to do anything except admire but Ian Glennie, who later played for Halifax Town and Forres Mechanics, was off his mark like a flash and kicked it. I can still see the muddy mark he left on the white flesh.

That was the high point of the season but it wasn't to last long. With both teams almost out on their feet, Rosehearty got the winner. Six–five, and a day we all remember – except Ian Glennie, who sadly died young.

19
Willingly to Schools

My mother's attitude to my education was a bit like her cooking – a case of good principles, expensive ingredients and indifferent practice. Well, Jean Allan's work as an educationist has been recognised as recently as 2006 by her inclusion among 830 women who lived between earliest times and 2004, in *The Biographical Dictionary of Scottish Women*, but that didn't mean that I learnt to spell. I blame it for the fact that I only made the connections between algebra and geometry when I had become a university teacher, and was working on a mathematical paper for publication in an academic journal. On the other hand, my education means that I can happily start a sentence with 'But' and a paragraph with 'And', and that I have only the vaguest notion of what an adverbial clause might be.

My Aunt Isobel, who married Mike Mackie and mothered six of my favourite cousins, once looked at me sadly and said, 'If you get your Highers it will be more by good luck than by good judgement.' I used to think that that was a terrible thing to say to a child. I still recognise that perhaps it was a little tough for I had no idea of what Highers were and was really alarmed by the thought that I might not get this obviously very important thing. But when, ten years later, I did indeed fail the school Biology exam which I needed to get into the university, and then failed it again, it was clear that Auntie Beldie's dismal prognosis hadn't been far from the truth.

John and Jean Allan believed in what was at that time known as 'progressive education'. To understand what that is you really need to know what it aimed to replace. The children sat then at little desks tight-packed in rows. They were not allowed to speak in the class unless spoken to by the teacher. If they wanted to ask a question they had to put a hand up and wait to be asked. They were not allowed to move from their desks during class except to go to the lavatory and then only if they could convince the teacher that their need was genuine. They had to call the teachers 'Miss' (even if they were Mrs) or 'Sir'. The classrooms were all brown and cream. There was no art, nor were there charts on the walls. The windows were high, so that the children couldn't see out to the sights of freedom. With classes typically of forty children each, there was little scope for individual teaching. The children had to read by turns round the class, chant their times tables and their 'gisintaes' ('Five gisintae twenty? Four. Five gisintae twenty-five? Five . . .') and they had to learn, not how to spell, but 'spellings'. The smallest misdemeanour was punishable by the strap, a hardened leather belt with two tails that nipped the skin of your wrist when brought thundering down on your hand. The strap was even used by some to help children understand the intricacies of the Three Rs – Reading, wRiting and aRithmetic.

Of course, a few teachers didn't use the belt and more didn't need it very often. It was at the training college in Dundee that a lecturer put the case for the belt. He gave each new class a demonstration. He put a line of bits of chalk on a desk and then shattered them in quick succession, each into a cloud of dust, with a colossal crash of his strap. The demonstration required no commentary. The implication was that, used properly, you didn't have to use the strap often.

Well, progressive education aimed to be just about the opposite of all that. Classes would have not more than thirty pupils. The children would learn much of what was required by playing rather than chanting. The alphabet would be out. The kids would soon learn to read in their curiosity to find things out. They would learn arithmetic by playing at shops. They would sit at tables in little groups and be allowed to chat about what they were doing and help one another. There would be woodwork, which would teach them about measuring, and lots of art, which would cover the gaily painted walls. The strap would be out. There would be 'answering back' which would be called 'discussing things' with the teachers, who would be known by their first names. They would learn about things by doing projects and finding out for themselves rather than being lectured about them in class. Children would be allowed to proceed at their own speed and they would have some freedom to choose which subjects to study on any one day.

I hope you will see that what the prog educationists were advocating in the 1940s is more or less what happens, in primary education at least, all over Britain nowadays. Prog educators have won the argument, though not everyone thinks they should have. Some think their victory has been too complete and almost no one thinks they were right in every particular. But explaining what went on at my latest strange school was one of the few banes of my childhood.

The first was a tiny one-teacher school in Alloa to which I went on the bus from Blairlogie with my friend Robin Kelsall. I know that it was run by a very nice lady called Mrs Grieve, who had trained under a leading light on the progressive movement, Montessori in Italy. I cannot remember anything about the school at all, which means I must have been not much over three. It was called Riverview and I have guessed that river must have been the Forth. Robin, being all of six, was put in charge of my safety for the eight-mile journey and luckily he published a book of memories of his childhood pals in Blairlogie, which has provided me with one anecdote from very early in my schooldays. My wife of forty-seven years loves this story, which doesn't show me in the sort of favourable light to which I aspire. In fact, it shows that I was an annoying bugger even then.

Now up to the time of my arrival Robin had been the clippie, Big Bella's, favourite. He had got to have some of Bella's unused tickets, to crank the handle of her ticket machine and, best of all, to play at drivers by standing at the front and using the radiator as a steering wheel. I threatened all that privilege. 'Charlie was a happy, chatty, curly-haired cherub who first awakened in me the sensation of jealousy.' No wonder, for Bella was quite taken in by this pretty little boy and she made Robin share his privileges with me. He described me smiling endearingly at all the lady passengers as I went down the bus. 'Charlie was a charmer but it didn't work on me. Things came to a head one hot summer's afternoon.' After school the two little boys had a quarter of an hour to catch the bus:

Even though responsibility for getting us safely to Bella's Hillfoots bus had been placed on my narrow shoulders, there was no way I was going to hold Charlie's hand. Children didn't have watches in those days but experience told me that we had to keep moving. Charlie didn't and wouldn't; he got slower and slower. I started to shout at him. His lower lip trembled and then collapsed. 'Ma wee leggies winna go ony faster.' I pointed out that if he wanted to miss the bus I didn't and, if he didn't get his wee leggies pumping, I was going to leave him. When we were quite close to our goal, in fact just round the corner, Charlie sat down, still wailing, and refused to go any farther. That was it and I was off. I hared round the corner just in time to see our vehicle slowly leaving its stance. Bella's three-bell emergency signal brought the movement to an immediate halt.

'Hurry up, Robin. You're late,' she shouted and then, 'where's wee Charlie?' By this time my lower lip was trembling and it too collapsed. Between sobs I blurted out that he wouldn't hurry and I'd left him round the corner.

'Get in there and sit down,' Bella's voice was rising ominously; wretchedly and hurriedly I obeyed. Through my tears I watched the small figure of Charlie appear round the corner, shirt-tail out, socks down to shoe-level, wee baggie dragging behind him.

'Come away ma wee lamb,' cooed Bella. 'What's Robin been doing to you?' Within minutes Charlie was chuckling away, smiling up at Bella with big eyes. My misery was now complete; would she tell on me? My gloomy view of the countryside was punctuated by the occasional self-pitying sob. Not far from the village Bella came and sat beside me.

'I'm no goin t'say anything about this, but if y'do it again, I will. Understand?' I nodded glumly. 'Good! It's forgotten then. He has.'

Charlie was kneeling on his seat, singing happily to himself as he gazed out of the window.

Blairlogie Boyhood, Robin A. Kelsall

Luckily I was too young to go to school when I was at North Ythsie, so I was not forced to go to the school where my mother's ideas about corporal punishment had been founded on the sight of her four-year-old brother going out on a regular basis to get six of the tag. But when we got to Methlick I said anxiously to her, 'I'm afraid the here school is a strap school.' She assured me that no one was going to belt me. She just didn't send me to school. But when 'Auld Tompie', a middle-aged and large, intimidating teacher who was no stranger to the strap, saw me puddling about in the river on her way home from school one day, she took a long look.

My mother was qualified to do it herself but for some reason she decided to get a tutor to come in and teach me and Donald Dewar (Scotland's future first First Minister), the son of a skin-specialist friend from Glasgow, who spent the summer with us. She bought us a curious two-seater bench which was attractively upholstered in a rich aristocratic red, but it was far too small. Goodness knows what her idea was, for while room was scarce in schoolrooms with forty bairns, there was plenty of it in the sitting room at Little Ardo. The two students jostled for room and leaned forward to write on the table that was far too low.

Sharing a small bench seat with Donald wasn't much fun. He was two years older than me for a start and maybe ten years further on in his schooling. While he was reading about the history of Scotland, I was just trying, without much success, to read. I remember nothing of the teacher and the experiment didn't last long. She either left or was sacked and I left Little Ardo for a while so that my mother could teach and I could attend Naemoor School at Rumbling Bridge near Crook of Devon in Perthshire. My mother and I lived in a tiny converted porch in the village of the Pool of Muckart. My father was still based at the farm but with all his radio work from Edinburgh and public-speaking engagements he came regularly and stayed for a night or two at a time. I went back for a look in 2007 and I can tell you that the outside dimensions of our house were fourteen feet by twenty. We had one room and a pantry with a sink and a cooker, and an outside closet. We took the school bus the two miles to Naemoor.

Our house may have been modest, but there was nothing modest about the school. It is now a Scripture Union activity centre called Lendrick Muir. It is a magnificent neoclassical sandstone mansion built in 1874 and now surrounded by a plethora of no doubt very useful but undistinguished low modern buildings. When I went to Naemoor School the low buildings were not there and it was a much more impressive place. Indeed the school didn't last long and I am sure that will have been, partly at least, because the huge cost of running such a palace could not be borne by perhaps forty students at a progressive school.

That is where my prog education started in earnest. There were two classes and, even at seven years old, I was in the younger class. It was taught by a really lovely person called Patricia Gilmour, who by pure coincidence had been among the evacuees to welcome my mother when she brought her son home from her own evacuation in Glenesk. She was tall and beautiful and had quite the sweetest nature. I never saw her annoyed with anyone in my year in her class and that was just as well, for she was not allowed any sort of corporal punishment.

In Patricia's class I proceeded to learn by playing and although I remember that it wasn't onerous, I can't remember learning much. I can however tell you about one of the prog tricks that were used to get knowledge and expertise into the children without forcing it in. It was designed to teach grammar. It was a sort of jigsaw puzzle in which there was a series of questions which you had to match with answers on the box. When you had finished you had to turn the jigsaw over. If you had got it right the picture on the other side made sense. If you had made a mistake the picture would be rubbish. That was fine, in some sort of theory, perhaps, but the young man from the farms of Aberdeenshire saw a flaw in the system. He noticed that the pieces were colour-coded. The verbs were all red, the adjectives were blue and the adverbs were mauve. So without trying to learn how to read the words, or learn the grammar attached to them, I could get the picture right every time. Soon I had all the puzzles in Miss Gilroy's class done and was going through to get the puzzles from the big class, which was taken by the headmaster, Mr Grieve. When I had done all the adverbial clauses, split infinitives and undistributed middles, I still hadn't learned a word of grammar. I am still not quite sure what a preposition is.

Then they would teach us about how to execute work to a plan by making us cultivate the garden. This was an excellent idea, especially as it got you out of the classroom and into something physical. Through this garden we were to learn to read by trying to understand the seed packets, to count by answering such questions as, how many potatoes are there in that bag? How many drills will we be able to get onto the bit of garden allocated? How many potatoes will we need to plant in each drill to use them all up? Mrs Grieve, who had been the teacher at Riverview, my first school, was in charge of the gardening class and noticed that, while some were digging and some were weeding and two were taking turns with the barrow, Charles didn't appear to be doing anything.

'Now Charles, what are you doing? You don't seem to be doing much.'

'Oh but I am,' I responded stoutly. 'I'm the grieve.'

With the patience which is a *sine qua non* of the prog teacher Mrs Grieve said, doubtfully, that that was very good but she'd like me to give someone else a shot at being grieve and for me to take a turn of the spade and of the barrow.

'Na, na,' said the aspiring grieve. 'They're feel enough, let them dea it.'

There are some that are born to lead.

One of the best features of Naemoor was the burn that ran past the school and the large ornamental pond that it fed. There were crested newts in the burn and tadpoles and we learned a good deal of biology from catching and looking after those.

But I had ambitions for the pond. We paddled in it but that wasn't enough. I was for making it into a swimming pool. I decided to have a fund and went for it. I pulled gooseberries from the garden of the house in whose porch we lived, used my sweetie coupons to acquire a pound of sugar and with some small help from my mother made two pounds of gooseberry jam. This I offered for sale in my swimming pool raffle.

It was a great move. Everyone thought I had been very public-spirited in donating a whole two weeks' sweetie coupons to the cause, but it didn't cost me much. There was, as part of the learning system at the school, a shop run by the children for the children. There we all spent our sweetie rations and somebody had the excellent idea of putting out a box into which people who were impressed by my magnificent generosity could put sweeties. So I ended up with far more sweeties than the ration would have allowed – and the credit.

Mind you, I got a terrible drop the day my raffle was drawn. Far from just one jar of my wonderful jam, there were heaps of prizes. There was a bottle of South African sherry, half a dozen eggs (a very valuable prize and practically unattainable in those days) and an oil painting of a disreputable looking miner who I thought looked like a Spanish onion seller. It was clear that my idea had been hi-jacked. People who were older than seven had taken over, expanded the raffle beyond recognition and kept me out of the loop. The young philanthropist was hurt but he consoled himself with all his sweeties.

And the swimming pool? I left before enough had been raised, the last I heard the fund had reached £16, and the school closed before a swimming pool was built there. The place is now called Lendrick Muir, and when I visited it in 2007 I could see there had been some sort of attempt. My wife Fiona and I found the pond and they had clearly had a go at making it suitable for swimming. I guess a digger had been in and deepened it by at least two feet. Sadly, it looks like the money ran out before they got the walls built up. The beautiful stone-built sides of the pond stop where they always did, at a depth of perhaps three feet. Below that there is just earth, which is gradually eroding away. It will all fall in soon. Still, I expect the crested newts are pleased. They don't like chlorine.

There were extensive grounds which we had the run of against the backdrop of the gentle Perthshire Hills. The only bit of our play I can imagine with any clarity is the housies we built in the woods and the battles we fought as Robin Hood or Robert the Bruce. It was really a heavenly place for a school and I think I was happy there.

A large part of that happiness was nothing to do with the school. My parents were friendly with Jack Martin, who had built up a small chain of shops selling upmarket fishing and shooting gear. I presume he liked me, because he taught me target shooting with an air rifle and fishing for trout in the tiny stream that runs past his Victorian villa at the foot of Glen Devon. Every day, it seems to me as I remember it, after the school bus had brought me home to the Pool of Muckhart, I got out my bike and cycled just less than two miles up the glen to the Martins' house. There I took out the beautiful six-foot greenheart rod the old man had given me, raked up some nice branlin worms from his compost heap and went fishing. Jack had shown me how to decide if a fish was big enough to keep. It had to be seven inches long and he showed me how he measured seven inches. It was from the tip of his index finger to his wrist. As I had a very small hand in those days, I caught lots of fish that were big enough for me to keep and I took them home to my mother. They were delicious grilled with butter.

Apart from that I had my loons in Glen Devon. I don't know how many we were but we played on the golf course where our greatest endeavour, which always came to nothing, was to catch rabbits. We guddled in the little stream where the fish were so small they didn't pass even my test of suitability for taking home. We also climbed the Seamag hill at the back of the village, right to the top. It amazes even me that my mother and the other boys' parents allowed us to do that without even any parental guidance, for it stands well over 1,400 feet high. My mother even encouraged me in a snowstorm once to go sledging alone up that hill, and I don't at all think she was trying to get rid of me.

It cannot have been an ideal arrangement for family life, although his diaries show that John Allan managed to visit us as often as once a week on his way to and from his various engagements at the BBC in Edinburgh or the Workers Education Association in Alloa, but for whatever reason, after a year or so we left Perthshire and went home to Little Ardo to stay. At least, my mother did. For me it was not that easy. I was now about eight and even she began to get serious about my learning to read. The Methlick School was still a strap school so I had to go somewhere else and there wasn't a big choice of prog schools available. Still, she found one that would do. It was at Culter, was called Lochnagar and was run by two very nice spinsters called the Misses Inkster. Jessie looked after the food and the housekeeping and Davina did the teaching. I went there during the week as a boarder, staying four nights in Culter and three nights a week at Little Ardo.

Lochnagar wasn't a very prog school. It had a very relaxed atmosphere with no strap. If you wanted to move, scratch yourself or even speak in class that wasn't a problem. But there was considerable emphasis on learning to read, write and do things like long division. We even had to learn our tables, though we were spared the gisintaes. At Lochnagar I did make a serious start on my schooling. There was plenty of room at Lochnagar, with a bit of hilly ground at the back which was called 'the Common'. Whether it was common ground or not we certainly used it as such for walks, play and bramble-picking.

There I had the only fight in my life that I can clearly remember. I was very surprised by it, and it was very short. This boy, who was a good pal, must have found me as annoying as Robin Kelsall had, for he came rushing at me with fists flailing. I flailed back with one lucky flail which hit him in the face. He started to cry and that was it. The other seminal moment at that school was when a member of the Deeside aristocracy invited me into the rhododendron bushes to compare 'front bottoms'. Most ungallantly, it seems sixty years later, I declined. I had no interest in hers. It never occurred to me then that she might have really been wanting to see mine. How selfish can you get?

Anyway, I was happy there but little did I realise that Lochnagar was only a holding operation. I was there to fill in time until my mother got her dream of a school of her own off the ground. It would put into practice the principles of progressive education and it would be a beacon for the education authorities in Aberdeen. They would welcome the trainee teachers to come and see how in learning by doing, a relaxed atmosphere and discipline could be achieved without the strap. They would make such a good job that all the children in the North-east would one day be taught in that way. The new school would take a bit of financing. So my parents took a partner, and that is a very poignant story. There is no doubt that my mother's view that a better way should and could be found was influenced heavily by what had happened at Tarves school to her little brother John. Little John was now six foot four and laird of the finest farm in the Howe of the Mearns. So a company was formed with Jean Allan, John Allan and John Mackie as directors. Very grand premises were found at 39 Albyn Place, one of the best streets in Aberdeen. It cost £7,775, had plenty of rooms for three classes, a residence for the caretaker and his family, and bedrooms for a few boarders. There was even a mews cottage at the foot of the garden which made a very convenient house for the senior teacher. It also had a very large garden for games and gardening. I was very proud.

A very clever and patient Ayrshire man called John Wilson was employed to teach the older children, who ranged to perhaps fourteen, and an absolute darling of a primary teacher called Annie Craig Scott from Perthshire was employed to look after the little ones. My mother's school pal Ella Anderson, by now a local minister's wife called Mrs Nesbit, taught the middle class. My mother was headmistress and did some teaching. We started with about forty-two children, most of whose parents paid fees. Some paid what they could, including Hamish Laurie, the artist, who paid one term's fees with an Impressionist watercolour of a castle on a hill, which my parents left me and which still stands above the mantel in my sitting room. I know that John Mackie had some more of Hamish's paintings so I guess those were fees also.

But who were the parents? Who would send their children to a school where they would get to do what they liked, where there was no strap to keep discipline and where according to the word on the street, they threw dinner plates at the teachers? Well I don't really know but some were of course teachers at schools and the university, who knew the principles of prog education and believed in them. One or two sent their children to St

Nicholas as a prophylactic against pollution by the working classes. More were attracted because they were at their wits' end because of bullying by other children or teachers in their corporation schools and were willing to try anything.

Of course, it just wasn't true that we got to do what we liked. And I never saw a plate thrown at a teacher. But there's aye some water far the stirkie droons. We got quite a lot of say in which subjects we studied at any one moment, though we had to study something and as most of the learning was done on your own or one-to-one with the teacher, there was no need to do exactly what everyone else was doing all the time. And no one on the staff would have made a terrible fuss if a plate had been thrown at them as long as no one was hurt. In fact the saintly Annie Craig Scott would go further than that. She had a very scared little new boy who wouldn't even reply when spoken to and seemed wholly miserable. She couldn't teach him anything because he wouldn't respond to her. One day she was discussing things with my mother and I heard her say, 'I am making some progress with James, I think. He kicked me today.' Her eyes shone. She was getting through his barrier of fear. I expect it was seen as another step forward when James broke a window in the school with a stone. He didn't get the thrashing you may feel he deserved but he was told, in no uncertain terms, that his mother would have to pay for the window. The headmistress heard him in the lavatory later that day speaking to himself. 'I don't care. I don't care if my mummy does have to pay for the window. Shite, shite, shite.' The inhibitions were disappearing fast. It's when I hear stories like that that I am glad I was not called upon to be a patient, progressive, primary teacher.

At St Nicholas we had a varied curriculum. We had the three Rs of course, but we also had Physics, Chemistry, Latin, French, Scottish country dancing and Eurythmics, which is a kind of less structured ballet. We had swimming once a week, football at Harlaw Park in winter and cricket in the summer. John Wilson took his whole class up Bennachie without a risk assessment or full disclosure. And my mother took all the older ones – those from ten to fourteen – to Bath in Somerset to the International Festival of Music and Drama. We learned a bit about eighteenth-century architecture and saw the iconic Royal Crescent, bathed in the hot springs, listened to Sir Thomas Beecham's symphony orchestra and heard Handel's *Water Music* performed on a barge on the River Avon. It was not an ordinary school.

In 1952 at thirteen the farmer's son from Little Ardo was sent to Devon for his secondary schooling. 'What on earth are you going away down there for, when Scottish education is the best in the world?' That was the question I had to try to answer a hundred times. The answer was quite easy really. My parents thought Dartington Hall and its headmaster W. B. Curry were the best in the country, if not in the world, and they had studied it. They may even have been right. Another factor was that the school was part of a great experiment by L. K. Elmhirst and his wife Dorothy, designed to revitalise this old estate. Instead of watching as populations drifted from the land to the towns, they would have industries in the country. They had a big building firm, a pottery, a cider press and a music school as well as the school which I went to. My parents were great admirers of the Elmhirsts and Leonard admired John R. Allan's work in broadcasting as well as his written work.

Certainly it was a wonderful place. The school was like a university really. We each had our own bed-sitting room. The subjects were taught in sequence between nine and six each day. You couldn't do them all but you could do any combination you liked and enjoy a free period when those which you had chosen not to study were being taught. That meant that during class times only about half of the pupils would have a class to go to. The other half were free to study in the library, play in the soundproofed music rooms, hang about, sleep or, in the case of the new kid from Scotland, play football on the lavish sports fields.

There was no religious instruction, no daily assembly, no rules about table manners and no dress code. Well, no official dress code anyway. The unofficial code was American blue jeans. Drainpipe trousers of the type I had only seen worn by snake-hipped cowboys like Alan Ladd, until I got to Dartington. How they all got them I don't know, for they could not be had in the ordinary shops, certainly not in Aberdeen.

My mother knew how important it was for young people not to feel left out in matters of dress. She had had problems with me before when she bought me a coat that buttoned on the wrong side and got me tormented for wearing girls' clothes. She had seen the Dartington children in their jeans when we went to view the school, and I heard her saying to my father, 'We'll need to get Charles some dungarees for school.' Even I knew that was not what everyone else would be wearing but I wasn't that worried.

Poor Mother, she didn't even manage to get blue denim dungarees. The nearest she could get was brown ones. So when I appeared among these fashion-conscious rich kids with brown dungers, complete with bib and rule pocket, but no belt, I was immediately dubbed 'the bricklayer's apprentice'.

The staff at that school were an interesting lot. Genralissimo Federico de l'Eglesia was the geography teacher. He had been one of the defeated government generals in the Spanish Civil War of 1936. The Biology teacher was Professor Marguerita da Camps, whose books on biology had been burned by Franco's thugs when they took over Madrid University, and her artist husband Guilliermo, who sold stone jewellery. There was Josie Warberg, a German refugee from Hitler who we found comically like a camp commandant. She was in charge of catering and had to keep order in the dining-room. 'Ruas, Ruas!' she would shout in her Germinglish 'get owse of ze dining zimmer'. Another German taught Latin. He was Ludwig von Rosenburg and had a glass eye which he used occasionally to turn in its socket, which made it even more difficult to concentrate on declining Latin verbs.

The gym teacher was a giant and the first man I ever saw in the flesh who had built up his body with weights. Tom Larsen had competed for Sweden in the 1936 Olympics and was a very fit sixteen stones when he taught me gymnastics, but he had had to work hard to achieve that. At the end of the war he was released from a Japanese prisoner-of-war camp weighing nine stone. I could go on and on for they were an odd lot, but I must mention Jarvis Crispin, the Chemistry teacher. He was nuts about cricket and it astonished me when he went down, alone, to the nets to practise bowling. He put a handkerchief down on the wicket at just the place where a good length straight ball should land. I didn't know anyone who could hit a bath towel so I was prepared to witness something right out of the *Boy's Own Annual* or 'Nick Smith; It's Runs that Count'. Well, I watched Jarvis for at least an hour and for quite a while on numerous other occasions, and he never hit the handkerchief once. Indeed, on that first afternoon, it took several bowls before he even managed to get one inside the net. Still, that's the Corinthian spirit.

The Dartington equivalent of 'the school where they throw plates at one another' was that 'they don't wear no clothes there'. As usual there was a dribble of water but not nearly enough to drown the stirkie. Bathing at the swimming pool and sunbathing on the flat roofs were normally done without clothes on. It seems sensible now and it seemed sensible then. The headmaster's view was that nudity, far from raising the sexual climate, actually lowered it. Goodness knows, but you don't need rocket science to tell if naked teenage boys are aroused, and in four years I never saw one.

Typical was the story of the group of ladies from the Women's Institute who arrived at the big double doors which led into the Quadrangle. They rang the bell and the door was opened by a child with no clothes on. 'Good God!' exclaimed the leader of the party.

The child said, matter-of-factly, 'There is no God,' and slammed the door.

Anyway, the kid from Aberdeenshire settled in fine in this rather odd place. I really enjoyed myself: there was loads of sport; the lessons went reasonably well; the girls were friendly but not too aggressive. I was really good at ballroom dancing to records of Victor Sylvester's dance band but never got the hang of jiving. There was no sort of ban on wandering about in the countryside and I found I got on very well with the Devonians. By the time I was fourteen I was playing regularly for Dartside Rovers, a men's football team in Totnes, the nearest town, and Johnny Gallon, our right half, was buying me a half-pint of beer shandy after the games. In my fourth and last year at school I was playing cricket every week for the Staverton and Dartington Estate teams and Johnny Gallon, who was a demon spin bowler in the summer, was buying me pints of Annie's mild beer. Aye, and that was the real stuff, kept in huge wooden barrels from which it dribbled warm out of the tap, once you removed the bung from the top to break the vacuum.

One aspect of the prog philosophy of education, of which I had experienced elements in my previous schools, was the democracy under which the staff laboured. At Dartington there was something called the School Meeting at which everyone, staff and pupils, had a vote, all attended, and all sorts of issues of discipline were discussed. No rules were made except by the School Meeting and the theory, which worked quite well, was that as the rules were made by the children they were more likely to be obeyed. One rule which was obeyed was that smoking would be allowed. But smoking was expensive even in 1953. Only one boy did smoke and it was thought that even he should desist while in class. Ah! But what about the staff? The meeting decided that the staff should not smoke in class if the pupils were not allowed to. Now John Harris, the French teacher, was a slender sparrow of a man, quick and nervous, and he told the meeting that they could say what they liked, he was going to smoke in class anyway. There was quite a row. It was settled after the meeting by the headmaster who took poor Harris aside and told him that if he wanted to work at Dartington he would have to work within the rules. The School Meeting was supreme and he must accept its verdict. He did, but he left soon after.

That was prog education in action.

All things considered, I did enjoy going to that strange school but it did cause me one moment of great embarrassement. It was when I came home for the first time from Devon. I was overjoyed to see the loons in the close and rushed out to greet them. Now, I had a pretty good ear. After only one term the Devonians were always saying they couldn't believe I was from Scotland, but the ability to 'talk' was no advantage with my loons. I told my tongue to say, 'Aye aye, boys. Fit like? Fa's team am I on, than.' But to my horror, what came out was more like, 'Hellowww, cheps. How aaahh you?' There was a stunned silence, which only lasted a few seconds. By the end of the game I was right back into the Doric.

The other and last thing of which I want to tell you is quite different. It provided a seminal episode in my education but I'd say it was anything but prog. By the time I was fifteen I had scraped through most of my O-Level exams and was desperate to leave school, to get away to university and what I perceived as the adult world. Two more years to get a university entrance was too long. I would do the courses in one year and be off. It was nonsense but they didn't, so far as I can remember, try to put me off despite a very average performance in class thus far.

So I found myself in a class of three with a very clever girl who was trying her entrance exams for Oxford and another boy who was a virtuous workaholic. Both were two years older than the teuchter from Aberdeenshire. Ted Fitch, the young, keen teacher, quickly spotted that Allan was a no-hoper – or was it his cunning bluff? Anyway Ted's method of teaching was to give us a talk about a subject, give us a reading list and tell us to go to the library and write an essay about the subject. And, there is no other way of describing it, he dealt with the good-looking, cockie sportsman by humiliating him. He did that by lecturing to the other two, jokes and all, and ignoring me completely. He made long criticisms of the others' work, showing them where they might have done better and lauding their strong points. My work was handed back without a word or even a look. I felt that so deeply that I made unusual efforts, read all of the reading lists, and put in essays that were far, far better. The first sign that I might be making progress came when he handed back the second improved piece of work. He had already spent five minutes each in detailed helpful comments on the others' work. When he handed over my piece on the dissolution of the English monasteries, he said, with just a hint of a smile, 'Quite un-Allanish'. I was left in no doubt that Allanish was not a desirable thing to be.

I credit Fitch with the fact that I passed the exam at the end of the year and eventually went to Aberdeen University, where I got first-class honours. But that was not prog education as I understand it. He should have encouraged me. He should have beamed with happiness when I improved at last. He should not have made me angry and he certainly should not have humiliated me – but it worked and I have always been grateful. I sometimes wonder if prog schools are successful because they attract teachers with enough smeddum to think things through for themselves.

I have said enough about the schools I went to but there is just one other thing. Even I, who was there, can hardly believe it looking back. Part of the prog philosophy is that children should not be made to compete. There should be no top of the class and no class prizes. Children should develop at their own speed and should not suffer the anxiety of not succeeding or the disappointment of failure. That worked quite well in the classrooms, although the teachers did put up the exam results and you didn't need to be very smart to see who had come top even if there wasn't a top. But what is non-competitive sport? Well, Tom Larsen was quite clear that it didn't exist and we had very competitive cricket, hockey and football teams which would do everything to win, short of getting caught.

To celebrate the coronation of Queen Elizabeth in 1953 they had a great sports day on the playing fields at Dartington. It wasn't just for the school, it was for the village, and indeed the games were open to anyone who was there. I entered just about everything. 'Tut Tut, how competitive! What's his problem?' I only remember one event specifically, which was the sheaf pitching. I remember it because the final was an all-adult affair apart from one lad in a brown boilersuit who came second, to wild applause. For that I was awarded a handsome gold-plated medal featuring the heads of the

Queen and Prince Philip. I still have it somewhere. But in fact I should have six gold medals. I won them and they were presented to me but the powers that were decided that I should give five of them away to other children who hadn't won anything. If you're one of those five lucky brats, can I have my medal back?

It is an interesting feature of my education that the four schools I attended have all closed. There is no Founders Day to which I can go and bore everyone with tales of my schooldays. There isn't even one Old School Tie let alone four that I can wear. It sounds like a disaster, doesn't it?

20
The Seasons' Return

There are years in which childhood experience is distilled and for me my father's first year back with us after the war was one of those. Perhaps it was that the dramatic weather we endured that year matched the changes John Allan began to put in place, as he put his stamp on the farm.

In January 1946, which is when John Allan's handwriting first appeared on the Little Ardo wages book, there was a staff of six to look after the single byreful of cows and their female followers: two hundred and thirty acres, of which about twenty-two were potatoes as part of a typical six-course rotation of three years grass, three years crops, which included turnips and black oats. The grieve was James Low, of course, who commanded no less than £17 for a lunar month, fully £4 a week, and a share of the farmhouse. For that he worked a sixty-five-hour week, though he was in charge and on call twenty-four hours every day. James Kelman, the dairy cattleman, had the same wage and another £7 11s 8d for Mrs Kelman's twice-daily darg, washing and steaming all the milking equipment. The Kelmans had a detached, two bedroomed cottage with kitchen and outside toilet. Then there was Bob Gray, a single man who had his own croft, Tillyfar, less than a mile away. Because he didn't need a house Bob had £1 in the month more than the two tractormen, Jake Clubb and George Cruickshank, who had £13 10s 2d and who each had a semi-detached cottage with running water but outside dry closets. The squad was completed by Bill Greig the orraman, who lived in the chaumer, the wooden shed for single men, which would have been called a bothy had he cooked for himself rather than being fed by Mrs Low. Greig was only paid £12 4s. When my father took over, none of that hard-working staff had ever had a holiday, though Kelman had every second Sunday off and the rest of the staff had every Sunday to themselves. Their contract was nine hours a day, six days a week, for a fifty-two-week year. All the married men were entitled to perquisites: four pints of milk a day, a ton of potatoes and a ton of coal and four hundredweights of oatmeal per year.

The thing to understand about farm work after the war was that there was no easy way to do things. There was the most tremendous amount of physical toil involved; hardly anything was mechanised. John Allan's earnest wish was to buy machines to make things easier for the men. He wanted to be able to pay the men more for less work. He wanted to cut the working day by shortening the hours. Cut the working week by cutting Saturday afternoon working and then eliminating Saturday working altogether, and finally by giving the men a whole week of holidays every year. He wanted every family to have a bathroom.

For my father the year started the day after harvest finished, though in truth there was no start and no finish to a farming year. It was and still is a continuous cycle of planting, sowing, reaping and selling. Still, in telling a story, you have to start somewhere. So the Hero's first whole year started on the eighth day of October 1946, when the last of the potato crop was lifted and stowed in the shed. It had been a struggle to get the potato pickers. It was not possible, as it had been before the war, to get a squad out from Aberdeen to sleep in the tattie shed and gather the crop. The shortage of labour after the war was so acute that everyone who had helped before the war now had steady jobs. But the farmer of Little Ardo managed to get the 'executive officer' (a leftover from the directed labour of the war years), to give permission to employ twenty schoolbairns and James Low had gone to the school to recruit them. With Mrs Rattray, a tractorman's wife, and 'Charles' (me), a squad of twenty-two lifted the twenty-two acres in eight days. For that the wage costs were 'Children, £3 7s 6d, Mrs Rattray £3 10s, Charles 15s'. My outrage at being thus exploited was explained, less than satisfactorily I thought, by the fact that I got my room and board for the eight days thrown in.

Everyone else was delighted except James Presly, whose farm of Wardford glowered across the Ythan valley at us, and still does. James was furious at John Allan's success in getting the schoolbairns. It is mentioned so often in Allan's day-by-day diary that Wardie's fury must have been real, but there is no explanation as to why he should have been annoyed, or whether the annoyance went beyond the typical inter-farm rivalry and the desire to be finished everything first. In 1949, when Wardie's crops had suffered drought and difficult harvesting conditions that had left quite a bit of his oat crop on the ground, John Allan even committed his antipathy to print. In his diary, he recorded the following naked prejudice: 'We assume that Wardie will be short of straw for the winter – and we assume it with a will.'

The day after the tatties were lifted the plough went in to the clean land left behind by the potatoes and the cycle started again. The ploughing was done with the two aged tractors: the old Grey Lady, the Ferguson with the three-point linkage which made it easy to turn at the endriggs, and the old Fordson with the bouncy seat, beloved of small boys on the farms. The old Fordie, which some of the men called Henry after Mr Ford himself, was a bit like a trampoline in its attractions for small boys. The tractors were a big step forward in themselves. Ploughing with horse had meant the man walking up to twenty miles a day up and down their fields following horse and plough. Now the tractormen could sit and smoke their pipes.

But it wasn't all good. The horseman had to work hard in controlling the depth of his plough and taking it out and steering it round at the endriggs, so he might be wet and his clothes might be frozen, but he was never cold. In the winter of 1946 the two tractormen were totally exposed to the weather; there was no work to keep up the body heat, and there was no cover from the weather. The nearest the mechanised ploughman came to protection was a plooin coat – a heavy greatcoat. Often that would be one of the trophies of war brought home by the demobilised troops. To make that aspect of the deal worse, the tractor could work as long as there was diesel and as long as there was light, whereas the horseman knew that when a horse had ploughed for five hours it needed a break, even if the farmers didn't think the men needed one.

At any rate, the ploughmen were pleased and the farmer felt he had taken a big step towards the mechanisation of the place, although it did take one of the iconic sights and sounds of the countryside away. The tractors roared up and down all winter. The patient team, the man singing or more often whistling and the pair plodding on, disappeared from the land forever.

The rest of the winter's work consisted mostly of feeding the cattle and clearing the previous year's crops from the place, making way for the next crop and turning oats and potatoes into cash. On a dairy farm much of the work was support for the dairy cattleman. He had to have a full barn of straw and hay, the loft had to be full of grain and the neep shed had to be filled with neeps or silage. It all needed so very much labour.

Take first the straw and the oats. The rucks were thrashed perhaps two to four at a thrash. This involved forking the sheaves, so carefully built at harvest time, onto carts where they were built again into neat loads. The loads were then taken to the barn, where they were forked again onto the table of the built-in Seggie thrashing mill. There James Low cut the string bands and fed the sheaves into the mill which separated the straw from seeds and chaff. The chaff had to be gathered in a hessian sheet and carried to the chaff house in the byre. The straw was removed by a two-pronged fork and stacked so as to fill the barn. When it was getting full this could involve two men, one lifting it up to about eight feet and the other building it from there to the roof. The grain was filled into two-and-a-quarter hundredweight bags and carried, on one man's back up the twenty-two steps to the loft.

We loons really liked a thrash. If you were very lucky you could get to dispose of the chaff, which you didn't get paid for, but it did make you feel like a man. But what we really liked was the rat hunts. These took place at the rucks. As sheaf by sheaf each ruck was removed, the rats who lived there found their world shrinking. They moved down and down into the bottom rings of sheaves. But of course there was only temporary respite to be had there. Eventually one would make a dash for it. Then another, then they would come in waves. We bashed them with shovels or spades, bits of wood or tried to hit them with whatever we had to hand. And the rats weren't stupid. One of my greatest memories in rat-hunting is when there was a breakout at one side of the ruck. Jimmy Low's dog Mirk, we three loons and a couple of the men chased after them. We were summoned back by frantic cries from my father, who had been overwhelmed by a break-out of rats who had spied that our defences on that side were now down to one well-dressed man. And there he was. He had a foot each on two rats and had caught one and was trying to catch another with his bare hands. Oh, the thrill of the chase!

We loons were glad the new farmer wasn't able to do away with the rat hunts for many years, but he made a start right away to sorting out the drudgery of the three-weekly thrashes. First, he got rid of the tremendous toil of taking the grain from the mill to the loft in bags. He put in a simple system of cups on a continuous belt which carried the grain up. When the cups reached the top and the belt turned down, the grain was spilt out onto the loft floor. If the grain was then needed it could be shovelled direct down into the bruiser or, if for sale, into a bagger. Well, trust me, that was a lot of darg saved.

Then to dispose of the chaff, he installed a series of ten-inch tin pipes. Along this piping the chaff was blown straight from the mill in either of two directions. It could be sent to the chaff shed, which was a boarded-up corner of the neep shed, or a shutter could be closed which would send the chaff straight to the old barn, which was now being used as a cattle court. It was an early example of automatic bedding and as such was much enjoyed by the staff.

But the labour-saving *pièces de resistances* were the elevators, which took the straw from the shakers of the mill up to and along the ceiling of the barn. There the straw dropped down from the first open hatch it came to. When the builders had enough straw through that hatch they shut it and opened the next one. From there whoever was building the straw into the barn had a relatively easy job. What had been hard work for two was now an easier job for one. The new farmer at Little Ardo liked that.

Cattle were fed on silage and neeps in the winter, and when John Allan arrived the silage-making went like this: on the first week in June the fields were cut with a mower. That left the grass in a fairly neat row. Along came the tractorman with a cart, and the orraman and another man filled it. They walked along behind the cart, taking a bout of lying grass each and gathered their row until they had an enormous heap which they then threw up onto the cart. There another man might build the grass systematically, which would mean more could be taken with each load. That was then taken to the pit, where it was tipped on top of the pile from which more graipwork was needed to spread it evenly to make sure there were no pockets of air. After that it was sprayed with treacle to help the fermentation which was the basis for the grass's preservation, and it was then tramped by a tractor running back and fore for several hours to squeeze the air out of it. In winter, the silage had to be taken from the pits in the fields where they had been harvested, to the neep shed. It started to ferment as soon as it was taken from the pit and exposed to the air, so it had to be brought in every three days at the most. The cows would not give so much milk on sour silage. This was again a case of back-breaking toil. First the earth that had been thrown onto the pit to seal it at harvest time had to be thrown off to reach the silage. Then there was no machine to cut the silage; that was done with a slicer very like a medieval beheading axe and probably out of the same era. Swinging this implement high above his head, the tractorman cut a line across the pit and then tore the silage graipful by graipful and threw it onto the cart till it was piled high. This stuff, which had been tramped by tractors, was quite solid and it took a tough man to take in silage for a yoking. It was then taken home to the neep shed and couped.

Then there were the neeps. Occasionally you might get the cattleman to take a shift at puin neeps but if he did that it was overtime. The neeps had to be brought home by the outside workers, especially the orraman. On a dairy farm that fed neeps to the cattle, the orraman could spend a huge proportion of the winter out in the fields puin neeps. Taking three drills at a time, he plunged his tapner into each neep in turn, whacked off the roots and then, in the act of adding it to the heap for collection, lop off the shaws. The three rows, all neatly topped and tailed, were thus laid together for collection. When the tractorman came to pick them up he would take one side and the orraman would take the other.

Handling neeps is all mechanised now but the farmer of Little Ardo in 1946 couldn't think how to make winter feeding easier on the men. He did in 1947 though. After the appalling winter, when for weeks on end they couldn't get out to the fields for neeps anyway, he decided to feed much more silage in future, and to get the pits close to the neep shed door. That way, all the carting home of fodder could be done in the balmy days of summer.

Clearing the sheds for next year's harvest meant getting the potatoes off to market – about 200 tons had been produced for seed for the English market. The new farmer was delighted with a crop that everyone agreed might have been as high as nine tons to the acre and in fact turned out to be over eight. Now that is a poor crop by the standards being set in the twenty-first century, when fifteen tonnes is nothing exceptional. But it was a good crop then, partly because they were deliberately trying to produce small potatoes for seed. The Englishmen were not fools altogether. They knew that they would get one new plant for every seed potato they planted so they didn't want big ones. As soon as the majority of the tubers were judged to be of seed size the crop was killed off with an acid spray. That stopped the plants from growing as well as ending for the year the danger of blight.

So it was one of the most important jobs as winter approached to get the potatoes dressed out of the heaps in which they had been stored. The 200 tonnes had to be graiped out of the heap and into a dresser. That was an electricity-driven series of moving webbed belts which allowed the earth to fall through and, by a series of different-sized riddles, sorted the tubers into seed and ware (fit only for eating). As the crop passed over the belt, between two and four people standing on either side picked out the damaged, rotten or misshapen potatoes into bags of brock for feeding to the cattle. The good stuff was sorted into one hundredweight bags and stored in a pile that grew and grew until wagons arrived at Arnage Station to collect them and take them away to England. The Little Ardo men were scathing about those little bags. 'Aye well, they're for the English market and they wouldna be fit for richt bags.' And the little bags meant that in 1946 the Little Ardo men had in the region of 3,000 of them to store at one side of the shed. Then they had to lift them by hand up onto a cart, where the tractorman would build them to a height of at least five feet. Then, when they got them to Arnage Station, they had to hand-lift them off the cart, carry them into the wagons and pack them to the roof.

This was what James Low was referring to when he told Maitland Mackie at Ellon feein market that he had a bad reputation because his men were 'ay gaun aboot wi' a sack o tatties on their back'. The new farmer at Little Ardo had heard that story and was keen to help. One of his first labour-saving innovations was the purchase in 1947, after great deliberation, of another elevator. A little petrol engine drove a continuous belt onto which the sack of potatoes could be laid and which would then convey them up onto the lorry. That saved the man on the ground punting the sack up onto the cart and it delivered it to the man who was building the load at above waist height so he was halfway to the highest he might have to lift it. It cost no less that £45 but the men were very proud of it even if it was a bit 'soft'. There was no elevator at Arnage Station, however, so there was nothing that could be done about the darg at that end.

Wagons were very scarce and that was a worry for the seed potato farmers. John Allan's diary has a rather forlorn entry on 30 October: 'Decided to dress so as to be ready for any wagons that appear.' Then Hallelujah, two days later it was: 'Got three wagons at Arnage. Must remember next year to order trucks in very good time, at Fyvie and Arnage.' Then followed a furious scramble to get 23 tonnes of seed away the six miles to Arnage. They had two 'Bert Grant' two-ton carts. Those had been a wonder in their day, holding twice what the old horse carts had held, but it seems a poor do compared to today's lorries which come right into the close, take away 25 tonnes at a time and take them all the way to the farm in England and even right out into the field there for planting. But this was 1947, so the two little tractors with two tons each, and a man riding shotgun on top of each load, ran back and fore to Arnage all Saturday. The men liked the job. They were getting on. They worked very hard for ten minutes at each end and then rode back and fore enjoying a good look at what all our neighbours were up to.

John Allan took a smug pride in getting his potatoes away so successfully. On 18 November he wrote in his diary, 'I understand my neighbours think I am getting too many trucks at Arnage. They know the way to Arnage just as well as I do.' He had been making a point of going down and speaking to the station master himself and in measured tones, instead of roaring abuse at him down the phone.

Everywhere he looked at the work of Little Ardo the new farmer was confronted by toil. Now he knew quite a lot about toil. He had done enough of it as a student to help pay his way through university to know what it felt like. And he had seen enough old men broken by it to convince him it was a bad thing. He had read enough about the majesty and virtue of toil to think of it not so much 'honest' as stupid. He claimed to have had a part in reducing the size of wheat bags from the two and quarter hundredweight killers to a mere one-and-a-half hundredweight in the early '50s. If there was a way of getting rid of honest toil, he was in favour of it. Hard labour might have been good for prisoners, though he doubted it, but for working men it was an unfair punishment for being poor.

Goodness knows there was still plenty of hard work with the graips left after the new elevator came home but that was just one of John Allan's innovations in his first full year in charge at the little farm on the hill. Silage-making was another area in which he found it possible to ease the burden on the staff at Little Ardo. He purchased what he called 'a silage pick-up'. This was trailed behind a tractor, which also pulled the cart. The grass was scooped up by a belt with spikes that stuck out, got under the grass and lifted it up and dropped it into the cart. To get a good load by this method you really needed a man in the cart to distribute it around the cart and tramp it down a bit while he was doing so. It was a bit clumsy, as after each load the silage pick-up had to be taken off and hitched to another tractor to allow the first cart to be emptied while the next was being filled. It was a slow business as there were no pick-up hitches in those days, but the pick-up saved a lot of hard graft. It also gave the farmer's son a very great fright. It was one of my greatest joys in 1947 to ride on the carts when the grass was coming cascading off the pick-up. I did a little good being in the cart as I was tramping it down a bit, but basically I was joy-riding. And the best bit was when we got back to the pit with a full load. The tractor had to be driven right up on top of the mound of silage, which could be eight feet high. I stayed on top of the load while it was being tipped. Up and up the front of the cart was tipped by the tractorman winding the big screw, until the load slipped off the back and I came tumbling down with the load onto the silage pit. That was very heaven. Until, that is, the day when I came tumbling down in such a way that I was quite buried in the grass. I didn't know if the men had seen that I was on the load. I struggled but couldn't move the heap of grass on top of me. I had visions of the tractor which would soon be tramping the pit running over and over me and tramping my lights out. I could hear the men talking. I couldn't hear what they were saying but they had no sound of urgency. Their next job was to level out the silage with their four-pronged graips. I could even feel the load on top of me getting lighter with each graipful. But would they stick their graips right into me to get a good graipful to spread farther over the pit? All was of course well. It was very curious, but when the men found me among the silage there was no nonsense of asking me if I was all right, taking me to the doctor to be checked out, given a thrashing or even telling me not to be such a stupid bugger next time. Sadly, I banned myself from that great game forever. I was seven and wanted one day to be eight.

Another of the great toils of the back end of any year on the farm was getting the muck out of the midden and onto the fields, where it could enhance the growth of the next year's crop.

Jimmy Kelman was in the byre before five every morning. There he had to feed the twenty pairs of cows tied by the neck in their stalls. That was a barrowful of silage or neeps. This needed the cooperation of the ladies, for the barrowful had to be got up between them to be couped in the trough they shared. Then he had to bring the draff hurley down the byre, giving a pailful of his own dairy mixture of draff (distillery waste), and oats often varied with flaked maize, bran or beet pulp and some fancy bought-in high protein dairy mixture. That was a pailful each; a big one for those that were milking heavily and less for those who were not doing so well. Then any beast who had sat in the wrong place and got dung on herself had to be scraped and washed and any udders that were dirty were given a preliminary wash.

The ladies all settled to their food, the mucking commenced. That started with taking any dung that had landed up in the stalls instead of directly out the back, down into the greip with a shovel. Then dung in the greip was pushed down the byre (which was built on a slope to facilitate mucking and washing), filled into the barrow and rowed out to the midden. That left the byre reasonably clean, but there was better to come. For the whole centre pass and the greip were then sweeled. I am not saying you would take your dinner off it but if anybody dropped his piece on Kelman's byre floor after it was sweeled he wouldn't have hesitated to eat it and would have made an effort to pick up the crumbs.

All that would have taken Kelman less than an hour. He was then ready to start milking. The udders were washed from a pail of hot water well laced with disinfectant, and dried with a towel. The milk units had to be attached to the teats, which could be a kittle job if it was a young heifer who didn't yet understand the score. The milk was then sucked out of the udder into pails by a pipe from the engine in the dairy. The full pails were carried down the byre and up three steps and emptied into the milk cooler at the dairy, at the bottom end of the byre. There Kelman had to fill the milk into ten-gallon cans ready for the milk lorry, and the dairy was also where the washing and steaming of everything that touched the milk was done by Mrs Kelman. Then the cows had to be bedded with straw and Kelman was ready for his breakfast. It was what John Allan called 'the cows' daily miracle – turning water into milk'.

After his breakfast, Kelman might manage a brief sleep by the fire before setting off again for the byre to see to feeding the calves, putting the bull in anywhere he was needed. That was an area in which Kelman had strong views. He hated the very idea of artificial insemination. He said it was a shame and he thought it was an abomination, that the bull with the bowler hat should be called to deny the beasts 'their pleasure'. He enjoyed the story of the crofter who, having failed to get his cow to settle took his neighbour's advice to try the AI. When the inseminator arrived, the crofter wasn't going to watch but he had everything ready. 'There's yer pail o hot water. There's yer towel. That's the coo. And there's a nail on the back o the door there, ye hang yer trousers on it.'

At about noon Kelman went home to his dinner, another cat-nap and, at three in the afternoon, all was to do again. So twice a day for the whole year the dairy cattleman had hand-mucked his forty cows in the byre. Even in summer, when most of the dung landed outside on the grass in the pastures, there were a few barrow loads after each milking. And of course in winter a large quantity of dung and soiled bedding were generated. All this muck had to be graiped into the barrow and rowed out of the side door, along the pass and into the midden. At first, when the midden was empty, that was an easy job. The midden floor was five feet below the tipping point. But as time went on and muck built up the contents would be well above the level of the gate to the midden. Then duckboards had to be put out and Jimmy Kelman had to manoeuvre his barrow along this to tip it. When the midden was nearing full he could have a precarious cobbly run for thirty yards on a series of duckboards in an attempt to find some fresh tipping space. When it was icy the duckboards could be treacherous and many's the time Kelman landed in his barrow when he slipped in the ice or let the barrow slip off the duckboards, where it came to an abrupt halt and he overran it.

That was toil enough, but it was spread out. At worst it could maybe be half a dozen barrowsful twice a day. It was nothing to the concentrated effort necessary when the midden had to be emptied in the backend. Again all had to be done with the hand graip. The midden was, as it is still, twenty-eight yards across by eighteen yards wide. It could be filled to a depth of five feet or more, so how many tons was that? Anyway graipful by graipful it had to be loaded onto carts to a depth again of perhaps five feet. This was then driven out to the fields that were to be ploughed and especially onto the potato ground. Then the tractorman tipped his cart up a little and drove slowly up and down while the orraman at the back pulled little heaps of muck off with a clique – a tool like a cross between a straw fork and a byre graip. It had a long handle and four or five prongs which stuck out at ninety degrees to the shaft. The clique fairly took hold of the dung and an orra-man could get the right amount to form a heap of the right size with one plunge of the clique. These heaps then had to be spread as evenly as possible so as to be ploughable. It was no good having heaps of straw that would snag the plough, so it had to be thoroughly shaken out in the spreading – again with the hand graip. It was unpopular and hard work and yet muck was still being spread at Little Ardo in that way as late as 1970.

But the farmer who took over after the war was able to make important progress in cutting the toil from mucking out the midden. Among his first improvements in 1946 was the purchase in November, at a cost of £335, of a brand new Fordson Major tractor. It didn't come with a backloader but shortly thereafter its rear-acting loader was able to do all the loading of carts with muck that was ever required at Little Ardo. That did away with many man days of the sort of sweat that romantics called honest, the men would have admitted was sweir if they hadn't been too proud, and the farmer called stupid. No one was sad to see the loader home to Little Ardo.

So those were the tasks that kept the men at Little Ardo busy through the back end of 1946. On 11 December they got the last of the potatoes away to Arnage Station. Every time the cattleman needed more straw there would be a thrash and there would be some grain to sell. And all the time the ploughmen took their chances to plough when the ground was neither too soft with the rain nor too hard with the frost. It was a case of getting ready for the sowing on the spring day and keeping the dairy well supplied. The farmer was well satisfied with how things were going on New Year's Day 1947. The new deep-digging plough that looked like it could rip the slates from the riggings of hell was doing its job and the ploughing was well forward, the muck had all been ploughed in, there was a month's turnips in the neep shed, the cornyard was still two-thirds full of stacks of unthrashed black oats – and the price was rising.

It was just as well they were prepared, for no one could remember a storm like the one that hit Little Ardo at the beginning of 1947. The new farmer was following his own advice and putting the land first. His diary of early December shows him ordering 100 tons of lime to sweeten up the Hill Park. When he discovered the government would pay half the cost, he doubled his order.

He was making a place for himself in this tradition – Bill Henderson, the very popular farmer of Little Ythsie of Tarves, took the farm of Old Craig in Udny, which had been farmed well for many years by Major James Keith of Pitmedden. Long after the changeover Bill Henderson grew a famous crop of turnips. Keith heard about the great crop and took the first chance to say to Henderson, 'I hear you have a very fine crop of turnips.'

'Excellent,' Old Bill agreed.

'Aye well,' said James Keith, 'just remember, I'm still standing in that field.'

It wasn't just that John Allan was improving the land and making the farm a better place to be fee-ed, he was becoming thirled to the place the way blackface ewes are thirled to their own hill and won't settle anywhere else. At least I think that's what this entry in his diary of 10 December is telling me. He had just been to call on Seggie, one of the giants of the Scottish grain trade, to see what he could learn about supplying what the trade wanted. They had got on famously. The grain trader had shown the budding farmer the Little Ardo sample and then a first-class sample and then explained the difference to him. Then he tried to sell my father some dear calf feed and been happy to be rejected. 'Seggie said all the right things about Charles. I didn't know his only son had been killed in the RAF. Why should we think our son will carry on our work as we plan it? If I couldn't give Little Ardo to Charles, I think of Jackie Large [a nephew] having it and it wouldn't be the same. Why should I think Charles would be good to Little Ardo? I feel now as if L.A. were the old family holding. It has taken the place of Bodachra. It's not wholly sentiment. I feel a sort of duty to the place.'

21
The Stormy Winter of 1947

The latest farmer of Little Ardo faced 1947 in optimistic mood. John Allan had made a good start to improving his wife's family farm. As well as labour-saving machinery bought, one of the hideous black wooden henhouses which had so blighted the east side of the house and steading had been removed. He had planted over two thousand trees where thirteen had stood before. He had cleared all the pens out of the piggery to adapt it to the modern need for lots of covered space that would be flexible. He wanted it for storing and dressing the potatoes and for machinery. He had installed a new water system, for the hygiene of a modern dairy, which required far more water than the old well was producing. The first of John Allan's plans for further improvement had been the cementing of that festering swamp, the farm close, and already he had bought a cement mixer at a cost of £86 10s, though the men would have hand-mixed the lot, and willingly.

Perhaps the fall of three inches of snow on the afternoon of 8 January was a warning of the storm to come. But it wasn't yet cold enough for snow to lie and to blow. The concreting of the close went on as the other work of the place allowed. On the 28 January 'Charles and I made a snowman which looked remarkably like James Durno of Uppermill.' On 1 February we got a set of chains for the wheels of the Austin 10. This was exactly that; a set of chains which went round the wheels so that the car was running on a series of chunks of chain rather than smooth rubber which polished the snow. With these fixed the driver was once again fooled into thinking he could drive anywhere without getting stuck. It grew colder and colder with always a bit more snow which didn't melt.

James Low and his squad had a considerable extra duty: keeping the roads open. That was important for any farm, but if you had a dairy it was vital. There was no refrigeration and the milk lorry had to get in to lift the hundred and fifty gallons or so, and leave the empty ten-gallon cans for the next day's milk.

Apart from the farmer continually getting stuck in the Loans, the best road to the farm, the only real problem was the milk. That was collected at eight o'clock each morning by Geordie Walker, the Mackies Aberdeen Dairy man who came with his four-wheeled lorry an hour after he had been to North Ythsie to pick up their milk and have his breakfast with the single men in the old kitchen, where I had once watched those lovely sausages fry, and lusted after them. In 1947 there were many days when the lorry couldn't get up the Loans. On such days James Low had to load the ten-gallon milk cans onto one of our two-ton carts and take them out to the end of the road to meet Mr Walker. Sometimes when the back road was too bad even for the old Fordson he would take the tractor through the fields to make his rendezvous.

Often conditions were so bad that the milk lorry couldn't even get to the end of our road and that was serious. The milk came steaming out of the cows and was passed through a cooler. That was just a series of pipes over which the milk was dribbled. In the pipes ran a continuous flow of gravity-fed well water. Groundwater, even in winter, wasn't that cold so the milk had a short shelf-life. We had just about enough milk cans for two and a half days' milk. That would be about fifty cans. So on the third day after the morning milking the cans had to get to the milk lorry and whoever took them had to be back with the empties ready for the afternoon milking.

And keeping our roads clear wasn't an easy job. It had to be done with shovels with which the men cut the snow into blocks and cast them out of the way. The roads were all narrow and had dykes at either side. When the wind blew across those dykes the road could fill in minutes. On 3 February Mary Low, a widow who came from the neighbouring Newseat of Ardo farm for milk when her cow was dry, said the mile of road she had walked was 'level frae dyke tae dyke', so she had had to take to the fields.

On 5 February, after the squad had cast the Loans once again, my father tied my sledge to the back of the old Austin 10 and, chains lashing the packed ice, drove me at what seemed like a furious pace down to the main road and back. The sledge had no brakes and luckily, even with the chains, the old car had little braking power either. My father assured me that I was quite safe. He was watching me in his mirror and if I looked like bumping into the car he just went faster. The Health and Safety Executive were nowhere to be seen, nor were they missed. It was a wonderful form of winter sport because it got round the problem from which the rest of our sledging suffered, that you got maybe half a minute of speed thrill down the hill and then maybe a quarter of an hour plodding back up again. Sadly my old man never did it again, though he got his chances, you may be sure.

We loons enjoyed the snow very much for the first few weeks. Every spare moment was spent sledging down the Howe, which runs the length of the west side of the farm. That was ideal for sledging as it was steep or very steep depending on where you chose for your piste, and there was no dyke at the bottom to crash into. The steep ascent of the other side slowed you down however fast you were going. I had brought with me from North Ythsie the big four-man sledge on which the young Australian airmen had taken me sledging during the war. It was the best sledge in deep snow but if we were on the roads sledging on packed ice there was nothing to beat Albert Low's little sledge with the narrow runners. She was called Fleein Flossie, and deserved her name. And we didn't only sledge on the Howe. We went down to Methlick and sledged down the main road from the Free Church to the village. Speeds on that run were very high and of course we had no lights. But the few cars there were did have lights, so we were all right.

There were epic snowball fights and when a thaw came we made giant snowmen. The technique we employed to get really huge balls of snow was to start them off on the top of the Howe. Only when we could roll them no farther, when they might be five feet in diameter, did we let them go down the hill. By the time our snowman came to rest on the bottom of the Howe in 1947 it was much bigger than we were.

But my greatest thrill in snow, apart from being tied onto the back of the car, was when I got my Aunt Mary's skis. With those I was able to negotiate a ski run that was the envy of the village. I started in the close at the farm. Down I skied past the piggery and through the open gate into the piggery park. From there it was a gentle slide down to the Howe. If I was still standing at the bottom of the Howe the momentum would take me up the other side, from which it was a gentle run down to the gate to the Millies field, the field next to Mill of Ardo. Back across Millies and through a gate and a fairly steep whiz down to the bottom of the Boddim Park and hope you could stop before you hit the dyke. It was wonderful and must have taken all of three minutes. Then all I had to do was shoulder my skis and in another half-hour or so I'd be taking off again.

On 7 February John Allan stuck his car at the cottar houses and plunged the 300 yards home on foot. The grieve had recovered the car by morning and when he said to him, 'I never thocht you could get hame,' John Allan's reply was, 'Drunk or sober, I aye get as near hame as I can.' So it seemed, for he was forever getting stuck that winter when wiser heads would have stayed where they were, avoiding travel at all costs.

On 8 February the milk lorry got stuck at the cottar houses and couldn't get in for the milk. The twelve cans had to be loaded onto a cart and taken up the road for transfer. Then the lorry had to be dug out and turned. It was no big deal for the men, but they would have thought more of it had they realised that the problem of getting the milk out would dominate their working lives for a good six weeks. That evening my parents had to abandon their car at the main road about a mile from home and walk. They soon discovered that it was easier in the fields. So much of the snow from the fields had blown off into the roads. My mother's first response when told she would have to walk was, 'Oh, my nylons.' Where so unfashionable a woman as my mother got nylons from in those days, and what she had been wearing them for, goodness knows.

On the 9th, the milk lorry got only as far as Woodlands, my grandfather's farm at Udny, ten miles away towards Aberdeen. It eventually made it to Keithfield, some four miles from Little Ardo, where our Jake Clubb met it with the Little Ardo milk and made the transfer. For the next few weeks the telephone proved its worth and the few remaining doubters among the farmers hurried to get connected. Walker had no mobile phone of course, but his progress was well reported from farmer to farmer.

There were two months of snow that year, and a large number of days when the roads were blocked. That was when the good people along the six miles of road from Methlick to New Deer all turned out with their shovels and cast the road, which was full to a depth of six feet in places. It was a wonderful community effort and there was a bit of a glow in the countryside the day it was finished and the first cars got through from New Deer. They might as well have stayed where they were. The north wind got up in the night and the whole lot was chock-a-block again the next morning. This time they just let nature take its course.

On a dairy farm like ours, the weather didn't make that much difference in winter. There was nothing growing. There was no winter crop to worry about or spray with stuff to protect it from disease. All right, movement around the farm wasn't helped by wreaths of snow which sometimes reached to the eves of the sheds, but everything was to hand. The cows were fed on silage and while some of the silage had been made in the field where it was grown, there was one silage pit just at the back of the byre. Straw was in the barn and oats were in the loft. And when those ran out James Low had a thrash and turned perhaps four rucks into straw to fill the barn and oats to fill the loft. He had a good stock of everything for winter: the fancy stuff to make up the dairy cows' rations, the diesel for the tractors; coal and sticks for the house had been laid in in measure suitable for the next Ice Age. The exception to that was that the milk had to be got out at virtually all costs and with the snow piled up those costs could be high.

On such a day James Low set out on such a mission. He had heard by telephone that the lorry could manage as far as Pitmedden, ten miles away. The roads were blocked for lorries but tractors, even with two and a half tonnes of milk on board, had a good chance of getting through, though it was by no means certain. James Low had a cunning plan. Instead of going by the main road he would go through the policies of Haddo House, the stately Adam-designed mansion which is the ancestral home of the Earls of Aberdeen. There the huge beech trees and the pine forests would mean that while there would be the same lying snow as anywhere else, there would be far less drifting. So James opened the gate at the Kelly Lodge and let himself in, being careful to shut the gate behind him.

The plan was working perfectly and James was almost enjoying himself chugging along at seven in the morning. 'These lazy buggers that work for the laird winna even be up yet,' he thought to himself and lit the pipe of

Bogey roll he had prepared for the journey. He was probably quite right about the estate workers, but sadly the laird was out and about enjoying the beauty and tranquillity of his inheritance. Well he might, for he had not long got back from six long years of fighting Hitler. He was not yet the Marquis of Aberdeen and Temair and so still called himself Major Gordon. The Major was not pleased when James appeared chugging along on his noisy, smelly little brute of a tractor.

The laird held up his hand. 'What do you think you're doing?'

'I'm on ma road to Pitmedden wi the milk,' said James fairly reasonably. 'I canna get roon be the road and wi the shelter o the trees this is far better.'

Major Gordon was a very reasonable man and would surely have understood but, perhaps to amuse himself for he could see that Jimmy was losing patience, he said haughtily, 'Don't you know this is private property?'

'Oh aye,' said Jimmy, his blood rising, 'foo's that?'

'Well,' said the future Lord Aberdeen, 'we were given this land by Robert the Bruce as a reward. We fought for this land.'

'Is it a fecht yer needin?' said James, roused, and making to take his jacket off. He wouldn't have cared that Major Gordon was nine inches taller than him and his feudal superior. Major Gordon might have been the captain of the London Scottish rugby team but he had never earned a pound by going three rounds with a professional at a feein market. Aberdeenshire was foremost a farming county and the milk had to get through. The laird backed down and the milk went through. I hear that the Major took a long lie for the next few mornings until the worst of the storm was past.

It really surprises me looking back on those days the number of times my father set out in the most unpromising of weather and, sure enough, got stuck again. He got used to ending up in a drift, draining the water off the old Austin 10 lest it freeze and crack the radiator and then walking home leaving James Low to mount a mission to retrieve the car as soon as might be. Now, John R. Allan had good excuses for setting out. He was getting a lot of work from the BBC at that time and was making a film called 'North-east Corner' for Films of Scotland. Most of the time he wasn't just joyriding, but what about the time he set out in the teeth of the storm to take me and two cousins to see a matinee performance of *Peter Pan* at His Majesty's in Aberdeen? He even stuck in the Loans on the way out from Little Ardo to the main road, but kept going. Surely it was saintly folly to carry on instead of calling the whole deal off, having a day by the fire and blaming the children's disappointment on the weather?

It was well worth it from the small boy's point of view. I was totally captivated by the action on the stage, although I could easily see the wires that allowed the cast to fly through the air. I did think it was a bit odd that two women, one of them called Peter, should want to set up house together in a tree, but when they appeared at the door of the Wendy house high up this tree, and took thirteen curtain calls at the end, I wanted it to go on for ever. I was ecstatic. But all that was nothing to the real-life drama of going home.

We left Aberdeen at a quarter past six on a wild night of wind that caused a lot of drifting. A couple of miles out we caught up with the service bus and sat in behind it till Potterton, which is six miles from Aberdeen. Then John Allan, impatient with the bus's speed, overtook it. That was a mistake. 'Thereafter I steered by faith, going through the casten [hand-cleared] parts as the wind blew up great clouds of snow. About Whitecairns [seven miles from the town] we saw a lorry on top of a dyke. Couldn't see road but didn't dare to stop. Jean put her head out of the window and gave directions. Charles rather worried in the back [rubbish! It was like the dodgems as we bashed from side to side of the great mounds of snow that were often well above the roof of the car. The passenger in the back didn't have the sense to be worried]. Twice I ran into deep snow at the side of the road but reversed immediately and managed to get out. Arrived at North Ythsie at half past seven and decided to stay the night. Seldom been so exhausted; though it was fun while it lasted.'

We had averaged about twelve miles an hour and got within eight miles of home. But I certainly didn't mind staying a night with my beloved grandparents. Then we rode home in triumph the next morning on the cart James had sent over with the Little Ardo milk. We sat with the empty milk cans. We were sure to get home but it was cold, though we would have been no warmer on the tractor, which had no cab. No doubt it was my mother's reaction to that trip that persuaded John Allan that his next improvement should be Duncan cabs for the two best tractors. The farmer waited to bring the car home once the snowplough had cleared the road to Methlick and got stuck once again in the Loans.

So, while the winter was a problem for the grieve, with the barn full of straw and forty cows keeping the byre cosy, there was nothing worse than usual about the job of the dairy cattleman at Little Ardo in the great storm. Mind you, the job was never a picnic. But it wasn't his work that made the winter difficult for Kelman, not even slipping and sliding his way along the duckboards of the midden, it was the commuting from his cottage to the byre. The two are almost 400 yards apart. For most of the eight weeks of the storm he couldn't walk on the road, which was at least three feet deep in snow all the way. The road passes by the garden of the farmhouse, and there the garden wall rises six feet above the road and above that is the hedge, which was a good four feet higher. And yet for much of the months of February and March the snow was right up and over the hedge . . . a depth of ten feet at least. So six times a day for Kelman, and four times a day for his wife, they had to take to the fields to get from home to work. And there were some wild times. On one occasion Jake Clubb set off at half past eleven to go home for his dinner, 300 yards in the other direction. He arrived back in the tractorshed having been stumbling about in a white-out for twenty-five minutes. His dinner was due north of the steading and yet the first building he came to was the piggery, about seventy yards south-west of the tractorshed. And, what made it worse was that the cattleman and his wife had to do at least three of their trips in the dark.

We loons felt so sorry for the Kelmans that we decided to clear the road all the way down to the single house on the brae where they lived. The six of us armed ourselves with shovels. Joe Low, by far the senior man at twelve years old, spaced us out and bade us 'cast our bit'. That meant cutting blocks of snow and throwing them to the side. The plan was not, of course, to clear the road, but to cut a single file path through the drift. My bit was only about three feet deep so I soon had a path two yards long. Then I saw that Albert had done better so I was able to find a spurt and before very long our paths met and we were able to move on to take on another ten yards or so each. It was great fun and we felt very important. When we came to the very deep stuff at the garden dyke we cast away until we were perhaps six feet deep, and then Jimmy Kelman Junior had the excellent idea that we shouldn't dig all this snow – we wouldn't be able to throw it out anyway – we should burrow through it. So we had a tunnel, twenty yards long. It was an excellent service to the dairyman and his wife. That afternoon they were able to walk in a civilised fashion up the road to their work while the successful snow-clearance squad watched proudly. There were two slight downsides. The first was that Mrs Kelman didn't fancy the tunnel much and Kelman tried to knock the roof in on grounds of safety, and secondly, the wind blew up again that night and, apart from the tunnel, which was still intact, all would have been to do the next day. But we didn't need to do it. The thaw came at last. Saturday, 29 March was the first day that the road into Little Ardo was open and stayed open. After that, spring with its rain was soon upon us and took away the snow, to everyone's relief – even we loons had had enough. We were anxious to get on with our football league and even we couldn't play football in deep snow.

Although there was still snow here and there in the parks – they even left a few bits of the last of the ploughing, skimming over bits that were too deep in snow and leaving it for a further thaw – they were sowing oats at Little Ardo on 9 April. It had been a long and at times furious storm. None forgot it and no one could remember a worse winter, though some of the old men tried hard. It was great to see the grass again and even the small boys were glad to be able to walk about without having to lift their feet high out of the snow to get progress. I remember well the pain in my soft young legs caused by hours of play in the deep drift.

The farmer was particularly happy at the thaw. He wrote in his diary, that first week in April, 'It is fine to go about the countryside after the snow. The birds are singing in chorus these last few days, instead of scattered tentative notes. There was a confident tuning-up in the hedge all afternoon; and now and then a hardy ensemble.'

But John Allan wasn't viewing all of the belated spring through rosy spectacles. In fact I am surprised at the deprecating tones in which he reported on the efforts of his neighbours to recover from the storm and catch up with the spring work. 'Clubb's ploughing Emslie's lea. It is nonsense we should be doing work for the crofters of Bennagowk at such a season. But if we don't do it who will? Our clients are a great nuisance. They rely on us and don't try to help themselves. They're horse farmers. Why don't they use the horse?'

It sounds a little bit like Marie Antoinette expecting the hungry French peasants to eat cake if there wasn't enough bread. But John Allan was developing himself a philosophy of farm work. 'Arnybogs has seventy acres to plough. Why should he have? Like so many others he loses himself in the back end'. That is a very important time. Get out the muck, get in the plough and plough as long as you can. James Low had all but six acres ploughed before Christmas. If we'd had all that still to do, how could we have spent the time necessary to get five acres of turnips up and away to Arnage? We couldn't, so ploughing in the back end meant £225 in the spring.' He had got 100 tons of surplus turnips sold at forty-five shillings a ton – 'It's so fantastic I can't believe it.' It was a good price if you think that forty-five shillings was more than half a week's wages for his top hand for each ton of neeps. On a farm with a staff of seven in 2008, that would be £200 a tonne.

But not everyone was pleased with the thaw. The break in the weather helped the vermin. 'Logan the old bull covered with lice and refuses to serve. I wouldn't either in that condition,' my father mused. Of course, in that condition, I don't suppose he would have been asked.

It wasn't just that John Allan was putting his stamp on Little Ardo with the grand cement close, which was so good for cycling and football, and the straw-handling system in the barn. The farm was also putting its own stamp on the journalist. Apart from taking a dim view of his neighbours' poor handling of the storm, was he also learning to blaw? James Kelman was generally recognised as a very good dairy cattleman and the performance of his cows was much better than most round about. Indeed, on 26 December, after a visit to his wife's relatives in the south, he was able to blaw in his diary. 'Our milk production compares very well indeed with yields in Angus. John [Mackie of the Bent, his brother-in-law] was really impressed by our average of three and a half gallons per cow; it is really nearer four gallons when you count what the calves get and the staff and ourselves. John's not much over two gallons and Milne's the same. John's butterfat average 4.18, Bob's butterfat 4.00, Little Ardo butterfat 3.9.' There is little doubt the better performance, with less butterfat, was partly because Little Ardo was further on in changing from Ayrshire to Friesian cows. But whatever, John Allan took considerable satisfaction from the Little Ardo cows' performance against what he saw in the fat lands of Strathmore and the Howe of the Mearns.

I am ashamed to say that when I read all that I thought my father was getting carried away. But then I read this in Maitland Mackie's diary. 'John R. [Allan] is getting 142 gallons of milk. We are only getting 162.' And Little Ardo had only one double byre while his father-in-law at North Ythsie had two.

There is no doubt the credit for the success of the Little Ardo dairy was shared. Maitland Mackie deserved some credit for building a wider, better-ventilated byre than the ones he had at North Ythsie. James Low ensured that the barn and the neep shed were always full. The farmer made sure there was always enough money in the bank to pay Kelman's wage of £17 a month and his bonus of almost six pounds a year. But the biggest credit must go to the way the dairy cattleman earned that bonus. He was a fairly angry man, who stood nothing that he considered was nonsense from his cows or his children, and kept a very clean byre. In the whole year there was never an instance of the dreaded *E. Coli* in the recorded milk samples and the harmless bacterial counts were always among the lowest in the area. The lady whose job it was on behalf of the consumer to ensure that Kelman produced clean milk was the milk recorder who came each month. She recorded the production of each cow and took away a sample from each for anyalysis. Miss Douglas was a very fierce public servant who James Low admired greatly because she 'stood in aboot till her work like a man'. When she showed Kelman how to wash the udders before milking he was glad it was the cow's udder she was scrubbing. Kelman hated Miss Douglas, but he was man enough to do exactly what she told him, and she knew her job.

Although the great storm was such a bother to the outside men, I don't know that it made much difference to Kelman in so far as the actual work was concerned. He was of course very worried about whether they'd get the milk out and whether they'd get him some empty cans for the next day's milk, but that was James Low's job. So was getting in enough neeps and silage and dairy meal to make the cows perform.

22
A Dry Summer

For the Little Ardo cows, who had been tied by the neck all winter, the great day of release was looming. The farmer wrote in his diary on 15 April: 'the neeps are getting tired and so are the cows'. Soon they would be out in the fields getting most of their own fresh food instead of chewing their way through preserved silage or hay. And on the 26th the chains were undone and out they went. The first few steps were tentative through the close, trying a few nibbles at the weeds by the side and the hawthorn hedge. Through the gate to the field and then, tails in the air, careering joyously down to the farthest fence, turning for a jousting session and tearing back up to the gate. It was what the farmer called, 'their annual athletics festival'. To see them frolic like that you might wonder that Jimmy Kelman managed to get them back again for milking.

There are two reasons why the cows co-operated, why they came tamely into the byre twice a day and allowed the cattleman to put on the chains that had held them all winter. One was that running around a field with up to three gallons of milk in your udder isn't much fun. And secondly, in the best conditions in the world, grass alone is a bit boring. The cows soon got to remember that the first thing they got after they landed in their stall was a pailful of Kelman's delicious dairy mash.

At the beginning of April then, the winter was over, the snow was gone, the ground was drying and the hash was on to get the crop in. The farmer of Little Ardo wrote in his diary: 'God bless the horse but keep him away from Little Ardo'. The tractors made catching-up possible. First was the oats. There was still the turnip ground to plough but by the 8th all but six of the acres which had grown grain or potatoes in the previous year had been ploughed. All that remained for the men to do was to break in the ground to provide a fine seedbed. This might have involved the disc harrows or the grubber on the lea to break up the clods or the land-levellers of which James Low was very proud. They were just three railway sleepers bolted together and trailed behind the tractor but they were especially effective where the ground was really dry. Then a straik of the light harrows and in with the drill machine. In 1947 there was no continual pestering of the land with serial sprays of manure, fungicides and herbicides. Ours was a combination drill so the manure was put on with the seed. Then it was a question of rolling the fields to press the loose soil round the seeds and encourage germination and that was it till harvest. 'And then shut the gate,' as James Low advised me when I was starting farming on my own account in 1973. It seemed to work in 1947.

James shut the gate on 17 April in 1947, perhaps a month late, and then it was more than time for the potatoes. As with the grain, the first thing that had to be done was to prepare a seedbed. But whereas the barley only required a flat, smooth field with perhaps the top three inches a free-flowing tilth, the potato needed as much black earth as there was. At Little Ardo that meant breaking in the top ten inches or more. That required working over the ploughed earth with the disc harrows and the spring tines. Then the deeper soil was torn up by one pass of the cultivators. To increase the workability of the soil and to help dry it out, drills were set up and left for a couple of days and then knocked down again – twice. Finally, when James Low was satisfied that the land was dry enough, the third set of drills were set up and the planting started.

Planting had finished on 23 April in 1946. The planting had been a delightful thing that year, done by a squad of schoolbairns direct out of the boxes in which the seed had spent the winter. The children walked along in the bottom of the drill dropping a potato then stepping over it and dropping another one by their toes. I used to wonder if people with smaller feet would be a good thing because they got more potatoes to the drill and therefore a bigger crop, or a bad thing because they used so much more seed. At any rate, the potato planting was a victim of my father's desire to put an end to toil. He didn't consider that the bairns would have given their eye-teeth for a bit of well-paid toil in spring.

A brand new potato-planting machine replaced the bairns in 1947. This was a belt at each side of a hopperful of potatoes. Then two people each side of the hopper sat on a plank and filled the potatoes into sections on the belt. As the tractor moved forward, the belts rotated, dropping the tubers into the drills. The spacing was variable, being set at ten inches, and there is no doubt it was a wonderful advance though I thought it a very cold business. The potatoes were all planted in six days and we finished on 17 May – because of the storm, a whole month later than in 1946.

On 21 May, only three weeks behind 1946, the sowing of the neeps was started. This had been a highly mechanised event even in the days of the horse. The tiny seeds were delivered in a thin trickle from a 'neep barra' onto the top of the drills. The standard issue did two drills at a pass but there was really no limit to how many seeders could be added as the barrow was not heavy.

The sowing of the neeps was finished on 27 May, in good time to start the silage on 2 June. That was bang on time, because the silage was always cut within a day or two of flowering, which is the time when the protein content of grass is at its highest, and that was always in the first week in June. The new pick-up worked a treat and the whole first cut, at between three and four tons an acre, was ensiled in pits at the back of the steading and handy for the neep shed door, in time for what was a truly life-changing event for the farm of Little Ardo.

On 12 June, with the crop all in, the silage made and a hiatus now until the neeps needed hoeing, James Low took his family to Greenock for a holiday at his brother's. It was his first-ever holiday. The first time since he was twelve years old that he had had two consecutive days on which he didn't have to rise before seven and go out to work. It must have been quite a shock to the system. Indeed, when the grieve came home his employer, filled with idealism for the promotion of the working man, suggested that in a few short years the holidays would be extended to two weeks. James Low was totally dismissive. 'Fit eese wad that be? Fa wid keep ye for a hale fortnicht?' It didn't occur to that good and willing man that he could go visiting for one week and then enjoy his bed, his garden, or the bowling green for a second week of leisure.

James Low returned on 17 June in good time for the hyowin (hoeing) to start on the 20th. The turnips were a difficult crop to grow. Even James couldn't just sow them and then shut the gate till harvest, like the oats. Turnips sown in the fourth week of May, or three weeks earlier, which would have been better, are germinating at the same time as a million other volunteers. And if the balmy days of June are suitable for the delicate turnips to germinate and grow, the same applies to all the other available seeds. Having made a perfect bed for turnips you had made a perfect bed for every weed known to Buchan. So great efforts had to be expended to reduce the competition.

The first step was mechanical. The sides and bottoms of the drills were scraped with a shim, which left most of the weeds in the bottom and only a small plateau undisturbed on top of the drill. Then perhaps a week later the hoeing began. The 'hyowin', sometimes called 'singling', had at least two objectives. The neeps were sown in a continuous stream, which ensured there were no gaps and gave the neepies a certain safety in numbers. But they could not possibly all have remained in the drills to maturity. So the first object of hyowin was to leave just a single plant every seven inches or so. In the course of singling the neeps, the hyower also got rid of all the weeds down into the bottom of the drill. It was an extraordinarily nice job singling the neeps. The plants might be a quarter of an inch apart, but they were often touching and you had either to push or pull the unwanted plants and leave only the single tiny seedlings on top of the drills. It sounds impossibly difficult, but in fact experienced hyowers worked at great speed, on most occasions one push or one pull was all that was needed to clear the width of the hoe and leave a single plant. If you missed, of course, you had to go back for another bash to separate a double. Occasionally you were too bold, or, if a stone got in the way, you could leave a gap, and there was damned little profit in gaps.

There was a certain etiquette about the hyow. The foreman went in first and set the pace. The second tractorman was next, then the third. He was followed by the orraman, and when Kelman or any other person appeared to earn a few shillings by taking 'a whilie at the hyow' he would come next. There was a nice roadman called Geordie Park who came occasionally. Posties and bakers who started in the middle of the night and finished before dinner time could bring their hoe for a while in the afternoons and we got the other tradesman from the village at the weekends. Last, where he could see everybody's work as he progressed up the drill, came the grieve. I didn't go to the hyow in 1947 but when I did seven years later, I came last. That was to allow me to fall behind without breaking up the main line of men who worked slowly across the field.

The entire squad, with the exception of the loon, strung out in a diagonal across the drills made a brave sight, marred a bit by any left-handers who had to reverse down the drills and spoiled the symmetry. So everyone was close together, all pushing and pulling at their hoes, and there was no noisy machinery to make conversation impossible. In fact the hyow was a great place for the clash o' the countryside. There were several strands to the conversations. There was gossip of course – heaven help anyone who was caught *in delicto* even a wee bit *flagrante* at about hyowin time. There was politics – in my first year at the hyow, it included merciless and seemingly interminable teasing of Jock Mackinnon, who had somehow managed to gain employment at Little Ardo despite not just being a Tory, but admitting it. Jock had been a corporal during the war and for a while an acting sergeant and was nicknamed Sarge. Anyway, Sarge had joined the Farm Servants' Union but he was not going to say he was Labour even for a quiet life. His torment went on for hours. Then there were bars. Those are now called 'dirty jokes'. They weren't nearly as dirty as what you get on the telly nowadays but the sad thing was that they weren't, as far as I could see even then, at all funny. In the interests of science only I offer an example. I tell it as near word for word as I heard it in the neep park in 1954: 'The doctor was looking at this wifie and he said til her, "If it wasn't for bein a man of the cloth I would like to take this examination further." The wifie says, "If it wisna for the cloth I'm wearin I wid let ye."' I am relieved to say I do not remember a burst of laughter at that bar. Then there were news-based jokes like the ones brought round by the postie about the time a local laird hit the headlines by coming out of one sex and into the other. 'Did ye hear there was a disturbance ahin Catto's henhoose?' 'No fit wis that?' 'They were looking for Dr Semple's cock.'

I wish I had taken notes at the hyow, for it was non-stop all day and there were often spells after suppertime. So there could be as many as twelve hours of uninterrupted chatter from a class of men who were known for their reserve. One topic I remember going on for most of one yoken (shift) and being reverted to several times in shorter bursts, was the size of Presly's Big Park. We were hyowin in the Piggery Park which looks across the valley at the fine farm of Wardford. Mr Presly's farm being on the southern slopes of Ythanvale was north-lying whereas Little Ardo, being on the north side of the valley, was south-facing. So poor James Presly had no hope of beating Little Ardo at being finished first with harvest, hyowin or anything. But one thing he did have was a park widely recognised as being forty acres in extent. And Little Ardo's men hated that. Some even denied it. 'Na, na. It's nae bigger than this een.' The others agreed that it certainly wasn't forty acres but they could see that it was bigger than anything at Little Ardo. Now John Allan had all the local maps and would gladly have settled the argument for them but no one asked him. The loon, who was quite glad that he was always half a lap behind the others and couldn't properly join in the debate, eventually fell heir to the maps and looked the problem up. The facts were, Little Ardo's Piggery Park: twenty-one acres, Wardie's Big Park: thirty-nine, and yet for the boys, the debate went on and on. It fairly shortened the hyowin.

The hyowin was finished on 2 July and on the next day the farmer and the grieve had an interesting discussion. John Allan had done a lot of sums showing that the buying of cattle to graze your grass in summer and the sale of them fat, or if they weren't fat, as stores when the grass went done, just couldn't match the returns he was getting from a mixed arable rotation. 'So why do they do it?' asked the farmer-economist.

The grieve-philosopher had the answer. 'There's mair fun traivellin through a park of nowt than sweatin in a park among hay.'

The harvest started on 14 August. The very first job was 'redding roads'. In order to let the binder get a start to go round and round the fields, the outside bout had to be cut by hand with a scythe. That was a laborious task and 1947 was the last time it was done at Little Ardo, so I'd better describe how it went. The scyther cut a bout between four and five feet across, leaving the straw in a neat line. The bandsman started by making a band. That was a question of taking a generous handful of straw and fixing it with one hand just below the heads of grain, then with the other taking half of the straw and twisting it round the other half so quickly that no one watching could learn what to do. He then laid his band down on the ground with the heads in the middle of the band, gathered a generous armful of straw, laid it on the band and pulled the band tight, twisting it again quicker than the eye could follow and then tucking the two ends in between the straw and the band. That makes a sheaf. A good scyther and a good bandsman could approach a mile an hour at that so it isn't as slow as you might think, but it was coarse work. To think that, in the days before the reapers and the binders, the whole harvest had to be taken that way.

The only benefit of redding roads was that it avoided the problem of cutting the outside of the field. To do that seemed to entail running the tractor along the top of the dyke. It seems blindingly obvious now that it has been spotted, but the alternative is to put the tractor to cut the first bout from the outside in. Instead of going clockwise, you could make the tractorman cut the first bout anticlockwise. The only downside would be slight damage to the standing crop from the tractor's wheels on that outside bout. Anyway, the end of redding roads was the next of Little Ardo's labour-saving innovations. But there was plenty of labour yet to be saved.

The binders went in on 14 August and turned standing crop into sheaves lying in rows. On 22 August it was all cut and the farmer of Little Ardo collected the cliack, the last sheaf to be cut and bound, and took it away to the house, as his grandfather and great-grandfather had done before him at Bodachra. He set it down beside his grandfather's grandfather clock in the hall. There it stayed until my mother got fed up of its gradual deterioration and the mice it attracted, and threw it out to the midden.

After the binder came the stooking, the setting of the sheaves up into stooks. That is sheaves leaning against each other with the grain up and so kept dry. At Little Ardo we worked in pairs at the stooking and, taking two bouts each, stooks of eight sheaves were made. Some people preferred to take five bouts and then the stookers took turns putting down the first pair of sheaves from the middle row and then added two pairs each from the two bouts at either side. Setting the stooks was important for they were built on a north–south axis. That way each side got an equal share of the drying sun. So all the Little Ardo stooks were set to point at the Prop of Ythsie. It was a good idea and I can quite understand why James Low was outraged at the farmer's son, when he started to stook, telling him that that wouldn't always be right. There would only be a certain line on the farm along which the Prop was due south.

In a difficult harvest the stooks would be blown over and soaked by the rains. Then they had to be rebuilt. There were mornings on end when the first job for the outside men was to go and set up stooks. Many's the young man who left Scotland for a new life in the empire thanked setting up wet stooks in October for pushing him to the point beyond endurance where he was able to cut his ties to the land and the Old Country. And in a really wet year, when the first signs of a drying wind appeared, it was quite normal to knock down the stooks so laboriously built to turn their bottoms up to the wind. But the harvest of 1947, though not a heavy one, was at least an easy one. The stooks only had to be set up once.

And so, on 27 August, we came to the next stage and I have been struggling on through the rest of this impatient to get to the leading. The origins of the term 'leading' come from leading the horse from the field to the cornyard with loads of sheaves. Getting the sheaves into the cart involved driving along the rows of stooks, pausing at each stook to allow loading. In the days of the horse this might be possible by calling to the horse or by making that coarse kissing sound which encourages a clever horse to move on or stop. But that was no use with the tractor. You needed a driver. Now on a flat park all that was needed was someone who could reach the clutch pedal and had enough sense to press it when a stook was reached and release it when the stook was loaded.

It is an exciting time, as witnessed by this peek at Methlick culture through John Allan's diary for 31 August. 'Davie Miller of Collynie went home drunk at dinnertime yesterday. He went to the grieve's house to tell him to lead in the afternoon and found the grieve dressing to go and look for another job. He took the grieve by the throat and the grieve threw him out of the house, breaking one of his legs. Davie is in hospital and the grieve is staying on till he is better.'

It was at the leading in 1947 that I first got a job on the farm that would otherwise have had to be done by a man. That is carefully to avoid saying that, at just days after my eighth birthday, I was fit for a man's work, but there was an opportunity for a boy who had not yet gone back from the summer holidays to his prog school in Perthshire, to be useful in the hairst park.

I cannot say that the squad were especially blessed by the inclusion of the farmer's son. I released the pedal too suddenly on many occasions, sending the poor tractorman, who had to build the loads as I had taken his cushie job of driving, sprawling among his sheaves. I didn't understand what he meant when he swore at me and said, not very unkindly, 'I doot ye're using Kangaroo diesel the day, Charlie min.' And several times I was jumped out of my skin by the roar of indignation from the real workers when I was so overcome by my importance and dreaming about pay day that I forgot altogether about taking my foot off the clutch. It was even worse when I forgot to press it again when we reached the next stook.

There was no doubting the young driver's joy and pride in his new position. But the humiliation suffered at four o'clock when the first of the bigger boys arrived from school was almost as great. I was told to get off my tractor, and when the lippy began to tremble, James Low made the thing no better by explaining that I would have to make way for 'the better man'. The better man was only twelve and I was there first.

My next job after driving the tractor between the stooks was building the cart loads. This involved keeping your balance on the cart for a start. Then you had to build one row right round the cart with the bottom of the sheaves outward and following that with a row with the grain outward to hold the outside row in place. After that you had to fill in the heart and start again. Six rows (we called them gyangs), or even eight, made a load, by which time the next cart was ready for you. It was hard work but it felt good to be getting one shilling and sixpence for every hour. In two days I could make £1.

That was of course nothing to the great day when I graduated to driving the old Fordson and the Albion binder, the one which had been converted from the horse days, with the great James Low sitting on the machine controlling the height of the cutting bar and the sails. That was superior work but paid the same as stooking. My stooking days also started with the grieve. I expect he chose to partner me so that he could set a pace that would make damned sure that I was tired at lousin time [the end of the shift], but not so tired that I would be unable to turn out the next day. My father was watching us as we started stooking in the byre park. It was a fine day and the grieve and I had just finished cutting and it made a daunting sight – twenty acres of sheaves all to be set up in stooks. I remember thinking it must be fine to be the farmer standing at the gate with a foot up on the second rail and smoking his Players Navy Cut fag. The watcher claimed to have overheard the following exchange between the grieve who had seen it all before and his student, who was eager about the whole process.

'By God Jimmy,' I said when we stopped for the grieve to take off his cap, scratch his head and fill his pipe. 'That stookin's fairly makin me tired.'

'Aye Charlie, I'm tired tae,' he replied. 'But it's nae the stookin that's makkin me tired. It's listenin tae your news.'

Now that I am older I have understood what he meant. There is sometimes virtue in not saying anything unless you have something to say. When I hear my granddaughters in full flow, I think I hear myself at the stookin all those years ago.

But I have got six years ahead of myself. The squad on the land was two forkers and the tractorman building the loads and the loon driving between the stooks. That sent a steady stream of sheaves to the cornyard, where the tractorman had to fork the load he had built on to the rucks built by James Low and Bob Gray, the orraman. Forking on the land was a fairly rough-and-ready business. No one cared very much how the sheaves got into the cart as long as they did so quickly. But in forking to the ruck the forker had to do a very nice job. The builders had to build each sheaf with its long side down and with its head pointing to the middle. And a builder like James Low needed every sheaf to land within inches of where it was to be placed and kneed in by the builders. There were lots of fights provoked by builders being shirty with forkers who retaliated by putting a sheaf exactly the wrong way up and so ruining the swing of the builder's work or *in extremis* hitting the builder full in the face with the stubble end of the sheaf and the builder jumping from the ruck to the cart for a bare-knuckle fight. It never happened at Little Ardo, but the forkers had to get it right.

The leading was finished on 4 September and Little Ardo, as we used to say, 'got winter'. It had taken exactly three weeks from the first cut to the arrival of the winter sheaf in the cornyard. It was just a fair cornyard, with about four-fifths of the rucks the farm had had in 1946. The short straw hadn't taken long to mature in the dry conditions and the crop, which had been so much later into the ground than in 1946, was three weeks earlier into the cornyard. All that remained was to get the rucks' lids tied on with strae rapes and thatched with a covering of the previous year's straw.

The potatoes also reflected the knock-on effect of the storms that had lasted into what should have been spring. They had been planted late and they had been short of water when they needed it most. The farmer feared he would have fewer than eight tons an acre, but he did just make it. Lifting started on 22 September with a squad of twenty Germans. They were prisoners-of-war from the camp at Monymusk. Though the war had been over for two years they were still there, and the farmers of the North-east were very glad of them. Most were from the Russian sector, the part that was to become East Germany. They were in no hurry to get home to where they would be dependent on the mercy of the Russians they so hated, and who so hated them. They were welcome at Little Ardo. James Low admired Germans as good workers, and if he was getting his tatties up without having to fight with the authorities to get the bairns off school, he was happy to let the war lie.

The Germans fulfilled the grieve's expectations about the quality of their work and the twenty-three acres were lifted in one week. So on the twenty-ninth day of September 1947 John Allan pronounced his first full year in charge of the family holding complete. It had been quite a year, but they had got through it with honour. Kelman had performed wonders with the cows. Throughout the storm the milk sent off to the dairy every day that it had been possible had always been more than in the previous year, and in the dry days of summer the little farm was often putting away twice as much milk as in the same months of 1946. A rough count suggested that the milk had produced a profit of a shilling a gallon, and the enterprise made a profit of £2,500 despite the £3,000 spent on capital improvements and the new machinery, not to mention the introduction of the half day and the week's holiday. Kelman's little cottage on the Big Brae had got electricity at last. Albert Low and I were playing down the brae that first night and the little house was fairly winking as the four children who were still at home took turns going out into the hall through to the front bedroom and clicking on the light. John Allan wrote in his diary a daft story about a lady in the Highlands who was looking forward to the coming of the power, and what she had been told: 'You press a switch and a light comes on. You press another switch and the fire comes on. And you press another switch and a hooer comes in and cleans your carpets.' I don't know whether the *v* was missed out on purpose or at what stage it was mislaid, but it wisna me.

Many of the farmers said the industry could not afford such extravagance but John R. Allan welcomed the holidays warmly. He wrote in his diary, 'I look forward to the day when the men can stand me no longer and I have to do all the work myself.' He was only joking, at least about doing the work himself. And he cannot really have imagined that his granddaughter's husband, who took over the farm in 1997, is in exactly the position he envisaged. The mechanisation which John Allan sought has made it possible for Neil Purdie to do all the work himself apart from certain jobs which are so highly mechanised and whose machinery is so expensive that contractors have to come in to do them. The farm workers have left and their cottages are let out on the open market. But John Allan welcomed sincerely the leisure the machinery had brought his men.

In the meantime that was how the farm worked. There would never be as bad a storm in winter nor as dry a growing season as that of 1947. But each year the seasons returned and each year small steps were taken to increase yields and reduce 'stupid' labour.

23
The North-east Gets Going Again

VE Day was of course best for those who realised what an escape we had made as a nation, but for one little boy who had only the vaguest idea of what it was all about, the day when victory in Europe was celebrated was pure magic. To be more specific, there was such a wonderful fireworks display in the village of Methlick that I have never been able to be impressed by fireworks since. I had a tiny room in the front of the farmhouse at Little Ardo. What had once been the upstairs landing had had a bit shut off to hide the stairs that led to the loft. There was enough room left after the stair was in to allow for a very slim bed and no floor space at all. I loved my 'little thin roomie'. I used to lie there on windy nights snug against the wind which rattled the old sash and casement window, and on 10 February after a great storm that blew the snow into great wreaths of up to nine feet, there was a small wreath in my little thin roomie, so hard had the wind blown against the leaky old single-glazed window. But on VE Day it gave me the best possible grandstand from which to see the fireworks in the village.

In truth, there were no fireworks. But on the Belmuir, one of the very few bits of ground left as open moor by the industrious folks of Buchan, there was, during the war and for many years thereafter, a Royal Observer Corps station. Somehow the plane-spotters had acquired a large surplus of flares. Half of the corps went down to the village and shot flares back to the station on the Belmuir where the rest of them retaliated with salvo after salvo of flares fired back at the village. It was as near as Methlick came to danger during the war. I don't suppose the Health and Safety Executive would have thought much of it but it was in the spirit of the time. And for one small boy with a grandstand view it was wonderful. Victory over Japan Day was tame by comparison. I remember walking back with my mother from a shopping trip to the village and seeing a few Union flags out but VJ Day was a poor thing. I hurried to my bed and sat for hours at the window but there wasn't even a squib.

Despite that disappointment I remember the end of the war as an exciting time. There was the return of the boys who were now men, and the race to make up all the investment in farming and in housing that could not be done in wartime. And there were all those things which had been postponed during the war years starting up again – such as the agricultural shows at New Deer and Ellon, and the sheepdog trials. Sheepdog trials? What sort of event is that to stage on the edge of Scotland's premier cattle country? Sheep, to the cattlemen of Buchan, are little better than hens (which are women's work), and other vermin. I remember my grandfather being reprimanded by his wife for sighing a bit too heavily when she said she was producing 'a nice lamb chop' for dinner. And a guest, not very embarrassed, said, 'Do you not like lamb Mr Mackie?'

'I used to like it for a change,' came the sour reply. I expect he had been given lamb already that month.

Perhaps Ernie Lee, the farmer who had made enough money planting tea in Sumatra to take from Lord Aberdeen the fine farm of Boolroad at Tarves, hadn't known about the distaste of farmers in Buchan and Formartine for sheep. At any rate, he was a man of great enthusiasms and was always up for anything. 'Auld Boolie', as we loons used to call him as he rode past in his light brown Ford Pilot, even went in for local politics and, better than that, put up the biggest by far (and perhaps the only ever) political banner in Tarves Square. It was just after the 1945 election in which Clement Attlee had swept Winston Churchill, our war hero, out of power and replaced him as prime minister. And Ernie Bevin had become foreign secretary saying, 'Give me a ton of coal and I'll give you a foreign policy.' So the banner appeared in the square at Tarves: 'ERNIE bevin for foreign affairs, clem attLEE for national affairs – ERNIE LEE for Tarves Affairs.' It was very unFormartine.

Ernie adopted the cause of building a new bowling green in Tarves and he put his enormous energy into the most ambitious plan for a gigantic sheepdog trial. The prize money was to be the biggest in the country. Shepherds were to bring their dogs from far and near. Major Gordon was to provide the grounds and Lord Aberdeen was to present the prizes. Ten thousand spectators were to come and produce a huge sum of money towards the green.

The prize money was there and the dogs came with their handlers, but the spectators were too few. My father wrote in his diary, 'Only 1,000 came. Poor Ernie. The bowling green fund must have lost hundreds.' Maitland Mackie, who was more particular about such figures in his weekly letter to his six children, was just as pessimistic about the financial outcome but put the crowd at 2,300. That seems like a good crowd for a sheep event in the cattle kingdom, but it was less than a quarter of the number for which Ernie had budgeted.

My mother and I were there and I remember that there was plenty of room everywhere except in the ice-cream tent. My mother refused to take me in there because you could catch 'infantile paralysis'(poliomyelitis) in such places. I wasn't convinced then and sixty years later I think it was a pretty poor excuse for denying a boy an ice cream.

Then Methlick had a sheepdog trials. That was a much less grand affair. It was held on the Auchincrieve field that runs down to the village and where you could get a reasonable view from the main street or from the Pleasure Park. It was only held the once and I only really remember two things about it. Just like in the old-fashioned ploughing matches, there was a prize for the furthest-travelled competitor, for the oldest, for the youngest and for the man with the most children. I don't know how much these minor prize-winners won but the champion shepherd won no less than £2. That would be half his weekly wage and so should equate to more than £100. It is just as well, for the decent man invited all his rivals into the Ythanview Hotel where he stood his hand. The round was twenty-six nips and that can't have left a lot to take home to the wife.

Tarves Flower Show got going in 1947. Maitland Mackie, my grandfather, had been a somewhat doubtful organiser of the show. However, he had done a great deal to make it possible, including driving the judges round to see the various classes of garden and trying to calm the excessive idealism of some of the committee. There is heavy sarcasm in this entry in his newsletter to his children:

> Last week I took the judges to thirty gardens before lunch. Still, I am now conversant with the merits of flower shows and gardening. It is an excellent thing – it is cultural; it is so good for the working man – better than going to a football match on a Saturday afternoon – it is an antidote to socialism – a buttress against communism. Jean may be right – I am a sooer aul mannie.

All I remember about it was that I entered the 'weeds' competition, as we called the wild flowers section. Most people just picked a few wild flowers and stuck them in a jam jar. Some got hold of a fancy vase and made wild floral arrangements, but I was much more scientific. I got a big piece of cardboard and taped my weeds on individually with the names of each written below them. The first prize was mine and I foresaw a lifetime of getting the annual congratulations of the village (and, much more importantly, of my granny) and two bob. But it was not to be. My idea of sticking the weeds on a board and naming them was such a success that in 1948 everybody did that and I was not even on the prize list. I felt I should have been allowed to patent my idea.

The Methlick Flower Show didn't get going again until 1949, when they combined it with a sports day with races on foot and on bicycles and a grand five-a-side football tournament. And in the pets' parade I was done again by those progressive thinking people who don't like you to be too competitive. I had adapted my hurlie to be a guinea-pig-mobile. I had painted it green with a little chicken wire fence round it and a heap of grass cuttings for bedding and put ten of my guinea pigs in it and trailed them round the ring. Well, it was no contest. With others merely carrying their kitten or tugging at their mother's dachshund, or with a budgie in its cage, I won hands down and got to lead the parade. But the victor was outraged. It wasn't that he thought one shilling wasn't a big enough prize. It was the fact that all the other competitors got a shilling too. At least when I won the boys' race (under ten) I got two shillings where the second only got one shilling and the third a sixpence. But again I was done out of it by the non-competitive competition organisers. When I presented myself for the under twelve boys' race I was banned. It was no use me saying that I was under twelve as surely as I was under ten. I had had a prize already so that was that.

But the big event was the return to village life of the Buchan League in 1946. The return of the boys who had become men in the war meant a sudden expansion of the number of men for whom physical fitness was more of the type that enhanced the playing of sports than of doing stupid work. So we had a really good and competitive football team. We six loons from Little Ardo never missed a match, where we cheered sarcastically when the opposing goalie took a goal kick and barracked the referee whether he deserved it or not. For some reason the refs were always provided by the away team so we got used to expecting refs to be against us. There was one ref who was almost spherical but who panted up and down pretty well, when I think back to him. We used to think it very witty indeed to say, 'At ref's sae fat, if ye pit a preen in him he'd burst.' And it didn't matter how often we said it, it was still as funny as ever.

It wasn't just the loons who thought the return of the Buchan League a great advance in civilisation. Even the farmer of Little Ardo used to watch some of the games from the gate to the field. Not quite in the crowd but there all the same. And he did a very unwise thing. When Methlick beat Tarves in the first round of the Brucklay Cup he said that if they got to the final he would take all us loons to the final in his Austin 10. Sure enough the Methlick team, most of them former soldiers, did make it to the final and the farmer had to take the six of us the whole six miles to the neutral venue of New Deer to play the might of New Pitsligo. There was a great crowd. From Methlick there came thirteen buses (counting double-deckers as two). There wouldn't have been many cars as so few had cars, but many many more travelled by bike, for everyone had a bike in the countryside in those days. The crowd would have been five deep all round the pitch and we loons were allowed to sit on the grass in front.

The Methlick team lines, had there been such a thing, would have read, A. Forrester, V. Foubister, D. Watson, A. Stephen, W. Geddes, A. Watson, A. Thompson, G. Nicholson, J. Clark, J. Taylor and P. Foubister. The result was, as far as John Allan was concerned, a perfect disaster. Methlick played well and were winning 3–2 when New Pitsligo were awarded a penalty. Despite all the derision we loons tried to heap on the kicker, he scored, so it was a draw. All was to do again five days later and, naturally, we loons expected our taxi driver to take us back to New Deer for the replay.

As I remember it, and I don't remember it that well, New Pitsligo went into the lead and were in front 2–1 at half time. Then John Taylor, our flying centre who had been relegated to inside right for the game, flew down the wing, cut inside and banged in the equaliser. Then followed a remarkable performance by John Clark, the tough fullback who had been promoted to centre to add some malice to the attack. They say that in one game when Methlick were being beaten 6–2 with five minutes to go he urged his men on with, 'Come on Methlick. It's onybody's game.' I will never forget the determination in his face in the great Brucklay Cup final as he absolutely barged his way down the middle and slammed the ball past the alarmed defence . . . and he did it twice. Methlick won and we loons went home with a contentment we have never forgotten.

The five heroes who are still alive will never forget it either. And the Ythanview Hotel in the village, where my mother and I got such a poor reception four years earlier, was the scene of such a celebration. With whisky at a shilling a nip and no food on offer, you can imagine it.

Another of the institutions of the North-east farming community which got going again after the war was the Highland Games. The Laird of Arnage restarted his Games with a huge crowd in attendance at the castle. I didn't get to that – after all it was a good six miles away – but I did get to the Gight Games, which were held beside the ruins of the castle which had been the family seat of the Gordons of Gight.

It was a great day. I remember looking down from the little farm on the hill into the vale where the Ythan wends its slow way to the sea. And the road to Gight was black with people walking and people on bicyles. I wish I could say there were some in gigs but I don't remember any. For the crowd of perhaps 5,000 it was bikes or 'Walker's Bus'. It was as though there were two rivers in Ythanvale, one carrying water to Newburgh and the North Sea and one carrying the good folk of Buchan and Formartine upstream to Gight Games.

All the best athletes of the day were there as well as the pipers and dancers. It was the first time that I can remember seeing a pipe band. The great George Clark was there. He was a wonderful athlete in his day, and this was his day. He won all the heavyweight throwing events except the putting which was won by Bob Shaw of Ballater, a much smaller man but who just seemed to have the hang of the putting. Clark even won the tossing the caber, which we loons sitting on the grass had difficulty in understanding. It seemed like for ages that the athletes took turns of trying and failing to turn the stick over. In the end, George Clark did manage, to an almighty cheer.

But George hadn't had it all his own way at Gight that day. He met his match in Stanley Ruxton, the Methlick bobby. George had arrived at the gate, where he was stopped by the secretary Charlie Reid of Stonehouse, the Gight blacksmith.

'That'll be half a crown.'

'Ye're nae chargin me! I'm George Clark.'

'I'm nae carin if you're the King of Siam. Aabody has to pay.'

'I'm nae the King of Siam. I'm George Clark of Grange and aa these folk have come to see me.'

Geordie Clark had a fearsome reputation. He was said to have killed a policeman, though people who said that never mentioned that it was in a car accident. It was certainly true that when he had been wrestling in America between the wars, the wife of his victim came up behind him between rounds and stabbed him in the back. But he carried on and won the bout.

Luckily, PC Ruxton, the Methlick bobby, was at hand. 'Now George, it's like this. There's a lot of money on the prize list today and the only way you can win it is to pay Charlie here his half crown.'

So George paid up and we loons from Little Ardo sat on the grass and marvelled. There were children's races and we loons must have joined in, though I don't remember them. In fact, apart from all those people and the caber tossing, I don't remember much about the great occasion. But I do remember that it was an important event. It was part of the return to normality after the war and the hall funds at Millbrex benefited to the tune of £128 17s 5d. That was a huge sum but despite that success, the Gight Games was both a restart and a finish, for they were never held again.

But the Little Ardo Games were. They were held the very next day on the farm and kept us amused for days. But what to use for things to throw?

Well, getting a caber was easy. We got Mrs Low's clothes pole. It was a bit too light but it still took a bit of getting the knack of keeping it upright until you had made your run and were ready to heave it over. Then for the fifty-six-pound-weight-over-the-bar we got a fourteen-pound weight, which was the handy weight of one eighth of a hundredweight in those pre-decimal days. That we threw over a rope held aloft on two ruck centres with nails chapped in at the various heights. We had a four-pound mash hammer with a long broom shaft for throwing and a smooth stone from

Mrs Low's garden for putting. And we didn't stop at the heavies. We did long-jumping, hop-step-and-jumping and even high-jumping. For that we made stands with nails chapped in to support the bar. And the bar came from a slatted henhouse floor which we found in the couples of the old stable. We kept breaking those slats but there always seemed to be another one on that old floor. We also had races, including the sack race, for which there was plenty of equipment on a farm producing seed tatties for the English market. In almost everything Joe Low, the grieve's second youngest, won, but I was usually second.

It wasn't just things getting going again after the war. A remarkable thing happened in our community that had had no precedent. The new laird, who had been Major David Gordon during the war and when his uncle died would become Lord Aberdeen, brought with him an English Rose as his wife and later Marchioness to all of us. He was a very popular, bluff, good-humoured fellow who had once captained the London Scottish rugby team and nearly played for Scotland, but she was the stunner. I said she was an English Rose, but she was proud to call herself a Huguenot, descended from the French Protestants who were driven out of France in the late seventeenth century. She was a classical musician, and a good one. At the end of hostilities the young couple started up a choir. How appropriate, you might think, to have jolly concerts performing bothy ballads in the centre of the greatest concentration of English-language folksong in the world. But it wasn't to be like that at all. The rehearsals and performances were to be in the Haddo House Hall. The members were to be the butchers, bakers, doctors and farmers from as near and far as they were willing to travel. Soon they would be engaging the principal singers of oratorio like Joan Sutherland (if there were any singers like Joan Sutherland) and the choir would provide the chorus. When they tackled the *St Matthew Passion*, Elsie Suddaby, Mary Jarred, William Parsons and Eric Green were the soloists. John Allan called them 'the archangels of oratorio in perpetual attendance on the Messiah'. Among the two famous oboe players in the all-professional orchestra was the best of them all, Leon Goosens, who came on many occasions to Haddo House. The Hammond organ specially trucked in from Glasgow for the event was sponsored by Boosey and Hawkes, the music publishers. Years later James Presly, the farmer of Wardford, told me with wonder still dancing in his eyes, 'I never dreamt I would sing in the Passion.'

It was maybe going a bit far, but the marchioness was later to describe Glyndebourne as the Haddo House of the south.

I had no idea that I was doing anything special but if we could see a photograph of the four choirs massed above those great stars you should be able to make out the Haddo House Junior Choral Society in the very top row, where we could touch the solid wooden rafters. And among them is a little fair-haired boy, in his kilt and still just pretty enough to be a girl. 'Oh lamb of God most Holy who on the cross didst languish,' I sang in my little-boy soprano and wondered at so many people being in the hall. The Health and Safety will make sure that it never happens again. And out in front was this tall, slender, beautiful woman conducting so elegantly with her long arms and slim, strong body swaying and teasing the music out of two hundred throats and from forty musicians. She was such a vital spark and she used to appeal directly to us to sing as though we cared, and particularly to smile. 'Look as though you are enjoying yourself – which you are!' she said in her slightly fruity contralto. All comers were accepted into the choir, whether they could hold a tune or not. Particularly tuneless people and those with voices like cement mixers were also welcomed but told to sing quietly 'because your voice is so strong'.

We children were gathered from the parishes that surrounded the estate. A bus went round on Saturday mornings and collected all who would go. None of the other loons from the farm went, for the very good reason that you had to choose between the choir and the Boy Scouts. And it was pretty well accepted round Little Ardo that you could never be much use about a farm unless you had been to the Scouts. Where else were you going to learn to tie a cairter's knot or get a knife with that special prong which would take stones out of horses' hooves?

So for choir I had a different gang, a gang with quines in it. There were about half a dozen of us who preferred to cycle the three miles from the village. I remember only two, and both were quines. There was Elsie Murray who was a dark-haired beauty and out-of-sight older, and Edith Chapman, who was way too beautiful and soon went off to Canada anyway. Edith had a tremendous mane of thick blonde curly hair which streamed out behind her. I will never forget her cycling away with no hands and saying ecstatically 'this is my hoaby' (which was how 'hobby' sounded in Edith's broad Doric). We were willing to sing about Christ's Passion but really we preferred singing Scottish songs or bothy songs. The distinction may baffle you, as of course bothy songs come from Scotland, but there is a distinction which is understood by people of the North-east. Scottish songs are those made up by Robert Burns or Lady Nairn and sung by Moira Anderson and Kenneth McKellar. Bothy songs are made up by people whose names are mostly forgotten and sung by people from the North-east who will soon be forgotten. Those are our folk songs and we liked them best. What a day it was therefore when in 1949 we all went on our bus to the Beechgrove House in Aberdeen and gave a concert which was broadcast live on Children's Hour on the BBC Home Service. I sang:

Fare ye weel ye Mormond Braes
Where oft times I've been cheery,
Fare ye weel ye Mormond Braes
It was there I lost my dearie.

Surely a song for someone who was old enough to have had a dearie, other than his mother, in the first place.

Apart from one appearance on the wireless, the highlight of my junior singing career was at the Aberdeen music festival, where I was entered in the Junior Scottish section. There was a choice of four songs, two for the boys and two for the girls. There were ninety-six competitors and I came second to the astonishment of my mother, which I found most unflattering. Why couldn't she have been like the other mothers and slated the judges for being tone deaf not to have put her little angel first? I think there were two reasons for my doing so well. One is that there were very few boys entered and the second was that nearly all the girls sang about a gallant weaver and made such a meal of the word 'gallant' as to make a fool of quite a good song. When I went up to sing 'Ca the Ewes tae the Knowes' the judges had just heard eighty pre-pubescent girls pleading for a GALLANT! weaver, whereas I sang as though I knew what a ewe was and where I might find a knowe for it. The other thing was that the judges remembered me. When my turn came the pianist asked me a question I had never heard before.

'What key do you sing in?'

'I don't know.'

The sympathetic pianist gave me a clue. 'Well. it's usually G or C.'

'OK. I'll have a bash at C.'

Sadly the beautiful Mrs Gordon, as she still was, took that success as a signal that I was ready to sing some serious songs. The next year she entered me for an English song category and I hated the song so much that I really couldn't sing it. I didn't need to know the key because I had to carry my music with me, but I couldn't learn the tune and I couldn't remember the words. All I can remember about it now, apart from my humiliation in front of the judges, breaking down and starting again several times, was 'Sir Marmaduke was an ancient Knight, good man, old man.' Well, what had that to do with a loonie from a wee farm on the southern margins of Buchan?

Another highlight of my days as a choirboy was singing the part of King Wenceslas in the carol service at the chapel at Haddo House in 1949. I was a big boy of ten then and my page was a very nice girl called Mary Rust from Barthol Chapel, but she was all of fifteen and towered over her monarch. 'In her master's steps she trod though the frost was cruel, but heat was in the very sod which the saint had trodden.' We made a very odd couple. I am afraid Miss Rust would have had very little shelter from her diminutive monarch, but everyone said it was wonderful.

At any rate the choir, which has now become the Haddo House Choral and Operatic Society and spawned the Haddo Arts Trust, has been an important part of Formartine ever since and I was proud to be part of it even if it did mean I never learned to tie a cairter's knot and therefore never earned James Low's complete respect.

I must tell you just a little more about Lady Aberdeen who, where music was concerned, always insisted on being called June Gordon. Her husband being in the army, she arrived at Haddo House from her parents' home in England to make a home for Major Gordon at the same time as my mother and I arrived to make a much more modest home for Captain Allan. I always felt I had something in common with her even if she was the marchioness, not just of Aberdeen but of Temair in Ireland. So I was very pleased in 1990 when I was asked to record her memoirs and sort them up into a booklet for sale in aid of her beloved choir. We got it done. I gave her the manuscript and I never heard any more about it. I offer you this anecdote, lest it all be lost.

In 1940 Major Gordon was with the Gordon Highlanders in the expeditionary force in France. Just before the Gordons were defeated and many captured at St Valéry before they had time to win at Dunkirk and the chance of a boat home, she received from him a valedictory letter in which he thanked her for being a loving wife and making their time together so good. He also rehearsed their dreams for starting a choir after the war – one of the many dreams which, as she had got this letter, would not be fulfilled. Written on the envelope was 'Picked up and posted by a French officer. Courage Madame!' Two weeks later she was back at her parents' home at Harrow, where her father was a housemaster, when the phone rang. The unmistakable voice of her husband, breaking all the security rules, greeted her and told her to meet him at the foot of the playing fields. She went and there he was, as right as rain. He had been sent home for some reason before the defeat at St Valéry and so had escaped all those hazards. The letter had only been written as a precaution and he had lost it. The two walked hand in hand home across the playing fields of Harrow and rehearsed again their plans for a choir, Haddo House and the large family with which they hoped they would fill it.

24
The End of 'Stupid Labour'

After the war the amount of toil on the farms really declined very quickly. The machinery, when attached to a tractor, could do just about anything. Little Ardo got a foreloader attachment home which in 1950 cleared 350 yards of snow from the Loans in five hours, worked by one man. The farmer estimated that it would have taken five men two whole days with shovels. And a digger was brought in to do some deep draining work. It made a hundred yards of drains in one yoken of five hours. I don't know how long it would have taken with spades, but it would certainly have run to man-weeks.

The potatoes really did away with themselves as well as the toil there anent. The twenty-three acres John Allan grew in 1946 were somehow consolidated into an allocation by the Potato Marketing Board of only fifteen acres, and those were not sufficient to justify the giant machinery that is now in general use among tattie growers. When I was a boy we used to say that they would never manage to mechanise lifting tatties, but it is all done by machines now. They used to go home to the shed loose in carts. The pickers filled their skulls and left them for the men who were 'teamin the skulls' to lift and tip into the carts. The modern way is for the potatoes to go straight from the digger into one-tonne boxes on the cart without anyone touching them. They are stored in those one-tonne boxes, and handling is all by industrial loaders with 'tippling heads', which allow them to be tipped into dressers or planters or whatever. So the 'stupid toil' is gone. But that only affected Little Ardo indirectly. Our acreage was just too small to bear the expense of all the necessary machinery so the toil was eliminated by our ceasing to grow the crop.

A muck-spreader was first tried in 1948. In February Little Ardo took delivery of a brand new Wilmot dung-spreader. According to my father, 'It threw up showers of shit spreading the dung far farther and more evenly and far quicker than a squad could have done with the graip. Little Johnny Kelman, anxious to see how it worked, stood right behind it and was well and truly top dressed with well rotted muck.' That was days of toil every year done away with a signature on a cheque.

John Allan tried small things to take the toil out of the dairy. He introduced a system of getting the milk from the milking cans to the cooler by pipe rather than by carrying the cans. An Allan Scythe with a snowplough attachment designed for clearing pavements was bought to do the sweeping of the muck down the byre. It was hardly used, as the dairymen found it easier just to get stuck in as ever with the brush. The sweeling of the byre with bucketsful of water with which we loons so liked to help and at which we became so expert, was done away with in April 1949, when a power-washer was added. But it was size that beat the Little Ardo dairy in the end. Year on year the yields went up but sadly the optimum size of the dairy unit crept up also. Forty cows may have been an optimum, just enough to keep one cattleman really busy, in 1946, but that number crept up year by year. By the mid-'50s Little Ardo would have needed to have a milking parlour and to bed the cows in courts and so do away with the twice-daily round of feeding, mucking and power-washing the cows' sleeping quarters. But the size of the farm was thought to be against doubling the herd. One byreful of forty cows just wasn't enough to justify the expense of a parlour and John Allan was fearful that they wouldn't be able to make enough winter keep for the eighty cows that might have been economic. Maybe he could have done it but by the 1960s he would have been in trouble again because the optimum size had by then risen to well over 100.

The drudgery of growing turnips was done away with on most farms by individual sowing by modern neep barrows, which did away with the need for singling and by cunning use of chemicals, but at Little Ardo the drudgery was abolished by doing away with the neep. After 1954 the cattle were fed only silage and yields continued to rise, if slowly.

And the silage handling was made much easier. Instead of splashing through the fields in the winter slush, the grass was brought home from the field in summertime and silage made in two concrete pits just opposite the neep shed door. Men would never again have to go out to the field in blin drift to keep the neep shed full in winter.

Not all the innovations were successful, of course. The grieve and the farmer were so impressed by the way the muck-spreader covered the fields with an even covering of dung that they thought to try it on the silage pits. With the benefit of hindsight I cannot imagine why they didn't think of loading the muck-spreader with clean earth, but they didn't. The juices from the muck seeped down through the silage all winter and, not at all surprisingly, the cows didn't like it one bit.

In December 1948 hydraulic tippers were ordered for the carts. Until then, every load of silage or neeps or stones gathered off the fields or potatoes had to be tipped by winding a very stiff handle for a very long time. At the front of each cart was a long screw over two inches thick. A metal handle just like the starting handle on an old-fashioned car had to be turned and that gradually tipped the front of the cart until the load slid out the back. Then, a much easier job, the screwing had to be reversed. But with the hydraulics in place it was just a question of shifting a lever and the cart shot up, shift it again and the cart slid down again.

One of the few occasions that time was scarce at Little Ardo was the spring when the grain, the potatoes, the neeps and any new grass all had to be sown. But that was helped in 1955 by the introduction of a park of winter wheat. The first year they sowed it before harvest and were very proud that they were getting on with the spring work before the previous autumn work had started. Not only did that spread the work but the winter wheat proved to be the best crop on the farm. They could only get it sown after grass so only one field was possible, but in 1956 they grew a parkful of wheat at Little Ardo which was so thick and that made so many stooks that the first cart had to reverse into the field for loading. At least that's what James Low told James Presly in the village. Compared to the crops of between seven and ten quarters of oats or barley grown after the war that was a crop which thrashed out fifty-two hundredweights or fully twenty quarters. It was a huge advance.

In 1945, when the Hero came home from the war, the output of Little

Ardo was the milk from forty cows, some corn and twenty acres of potatoes. Apart from a few cull cows and a few bull calves that was still the same in 1957, but that was a defining year. It was the year in which you could say at last that the hard physical toil, the toil that bent men's backs, was finally over at the little farm on the hill. Men might work long hours, but they seldom needed to sweat because all the routine jobs about the place were mechanised. The wages of the men had risen to the point where the farmer could say with pride, 'the most expensive conveyor nowadays is men's backs'.

The final step towards the elimination of 'stupid work' was taken in 1957, when the first contractor came with his combine to cut the barley. John Allan at Little Ardo had been twelve years behind his brothers-in-law. He just didn't have the acres to justify the expense of a machine which might only work for ten days a year. But by the late '50s the contractors were getting going with combines. Those great machines cut out the need for redding roads, the binding, the stooking, the leading and the ruck building, not to mention the continuous thrashing of the grain all winter. Best of all, the contractors cut out the need for the small farmer to find a lot of capital for his machinery. Harvest, which had taken anything from three weeks to three months, could be done in a week. In one pass of the combine you got the cliack sheaf, the winter sheaf and the year's thrashing.

Agricultural toil came to a sudden end, and I saw it. The contractor's combine was humming round the Lotties, the field in which James Low had shown my father 'how we fork sheaves at Little Ardo' in the summer of 1946. And there, the picture of contentment in the corner of the field was James Low, Meerschaum pipe well stoked with Bogey roll, waiting with his tractor and the cart with the high sides. He was waiting for the wave from the combine driver to come and run alongside and receive a load of the cut grain for dumping in the barn for bagging. 'By God,' he said to me with fierce relish, 'this is the way tae hairst.' Little did he think then that a couple of years later even the wave from the combine driver would be been done away with in favour of a flashing light.

These writings have been about a family, and a farm which has now been their place for 170 years. That peasant connection is still secure. John Allan's granddaughter and her husband, who are not yet fifty, are now in the farmhouse and they have two daughters who show every sign of being breeders, and some of loving the place.

I will finish with the farmer who supervised those great post-war changes. John Allan took over a farm on which the work practices had hardly changed and living standards hardly risen in 100 years. He had seen the working week reduced from sixty-five hours to forty-five hours, Saturday working done away with, and two weeks' holiday where no holidays had been before. He had seen bathrooms installed in all the farm cottages, a fine cement close laid and three thousand trees planted where there had been thirteen in 1945. He had seen his great grieve, James Low, bring home his first motor car, a maroon Austin Seven. Some said Low had the first personalised number plate in the North-east when his second car's number was JLB, for James Low's Bus, but that was of course pure coincidence. Better than all that, John R. Allan had written the *North-east Lowlands of Scotland*, surely the best book ever written on the culture of rural Scotland. James Low did the North-east a favour when, in the Lotties in 1946, he told his employer to take his boots off and let them that could work, do the work.

And it was not just the farm for which those first few years after the war changed everything. John Allan himself enjoyed a change in his personal circumstances which only he could ever have truly assessed and sadly he never did. The life-changing year for the farmer of Little Ardo was 1950. On 12 May the entry in his diary read simply and only: 'Got a little sister'. But that signified the end of a deep longing which had been a sore drag on his life. John Allan had never blamed his mother for leaving him and moving to Canada. He had made no attempt to reach her. He knew his mother knew all about him but there was never a sign, despite his Aunt Susie's best efforts in sending her his books, and his own radio broadcasts to the post-war Empire. The only comment I can remember him making on his lack of a relationship with his mother was: 'She was just a lassie', and of his books, the only sign of bitterness: 'I expect they went straight in the fire.'

But then suddenly Fanny died in Canada. His Aunt Susie wrote immediately to John's two brothers and his sister and told them they had a half-brother in Scotland. It was a complete surprise to them, even a shock perhaps, but if there is a God he will surely have rewarded them for their reaction. His sister Fanny caught the first plane for Scotland. She brought her mother's gold watch to be a memento of the mother who had never acknowledged John Allan in any way. I was not privy to their meeting, and any excesses of emotion were kept from my eyes, but I have never seen a man so happy as my father was in the weeks that followed. He took Prairie Flower, as he called her in his diary, to Westertown of Rothienorman, where she expressed disapproval of the batteries in which Mike Mackie was keeping hens, and after five weeks he took Prairie Flower to Glasgow on her way home to Canada. So far as I know, that was all he who wrote so much wrote about their meeting.

The following year he went out to Canada, discovered his two brothers and visited his uncles, the dreaded Chae, whose pigs he had established could not be relied on to land on their feet, and the wicked Tom, who sadly was no longer with the Indian Princess. Both seemed like benevolent and even respectable old gentlemen. The farmer of Little Ardo sent us holiday snaps of himself, able suddenly to keep up with his wife, with a squad of little nephews and nieces.

It was a great time and part of what he was writing about when he wrote his valedictory letter to his aunt and protector, Susie Rennie, in 1980:

> My dearest Susie,
> Life goes on but I am tired. The old bones are not very willing and I try not to push them too hard. And I am becoming forgetful. I forget to do what I should do. I forget the people with whom I went to university and those with whom I learned the trade of a journalist at the *Glasgow Herald* . . . and so it goes on . . .
>
> But you I cannot forget. The old people at Bodachra were wonderful to me. But your kindness to me was a source of strength then and has been ever since. I remember you best when you were a beautiful young lady and I was the little boy you so often gave tasty bites that you were cooking. And I remember your courage in saving me in the close from Chae. I don't blame Chae altogether. He was jealous of the affection I got from my grandparents which his hard work about the place entitled him to but which he never got in full measure. But he did used to torment me in a hard way. Then you took my side, you were my protector.
>
> I forget so much Susie, but your kindness I will never forget, till the last hour of the last day.
> With much love, John

The last hour of John R. Allan's last day came in 1986, when he was in his eightieth year. Jean lived on for another four years and died when she was eighty-two. By that time I was firmly established as the farmer of Little Ardo and, in John Allan's phrase, 'growing into my grandfather's clothes'.

Glossary

aa all

arles small payment in advance of wages paid to seal a bargain of employment

arnut poor man's truffles dug up on mossie banks – the organ of vegetative reproduction of a small plant with leaves like a carrot and delicate white flowers

ava at all

bar filthy joke

barra wheelbarrow

ben (the hoose) through, at the other end of

birdie's pizz birds' peas – the leguminous fruit of the vetch

blaw boast

blin drift white-out (snow storm)

Bogey roll rank pipe tobacco

bone davey horsedrawn implement for sowing bone manure

bout(s) pass of the scythe, reaper, binder or combined harvester

branlin (worms) worms with bright-red segments which are very attractive to brown trout

bree surplus liquid. Potato bree is the water potatoes have been boiled in. Midden bree is the foul liquid that seeps out of a midden and silage bree is the juice that runs out of a silage pit. Burns called whisky 'barley bree'.

breem broom

breem dog tool for pulling broom out by the roots

breem cutters	tool for cutting broom stems
breid	oatcakes
brock	part-rotten or chipped potatoes for feeding to cattle or pigs
brose	porridge made without cooking. Boiling water is poured on the oatmeal and then stirred up with the handle of the spoon.
buchts	handling pens for sheep
cairt; cairtin	a cart; using a cart to transport anything
caw	work with, operate
chaumer	single men's quarters where the men are fed in the farmhouse, unlike a bothy, where the men cook for themselves
cliack	the last sheaf cut in a harvest
close	farmyard
coppit	caught
corter	one eighth of a girdle of oatcakes
cottar; cottared	person living in a tied farm cottage; established in a cottar house
coup	tip. Coup a cart means empty a cart by tipping it up so the load slides out
couple	rafter
darg	hard work
diets	meals
draff hurley	metal barrow in which the dairy cows' fancy diets, whose base component was draff (distillery waste) were mixed
droon	drown
dubbs	mud
dunnock	hedge sparrow
eese; fit eese	use; what use
endrigg	turning area at each end of a cultivated field
eynoo	just now
fa	who
fee	*v.* to make a contract; *n.* payment for working for a year or six months
feel	fool, foolish
fitba	football
foo	how

fou	drunk
fraise	stroke
fut, fit	what or which
gaun tae the furth	go outside, for example to the lavatory
gey	very, rather
girdle	steel hot plate heated by suspending over an open fire; griddle
girnal	wooden box for storing oatmeal
gorbled	fertilised (egg)
graip	pronged hand tool for loading muck, silage or turnips (byre graip) or potatoes (tattie graip), or for digging in a garden (garden graip)
green lintie	linnet
greip	open drain in the byre in which with luck all the urine and dung lands
grieve	working farm manager. Like a foreman in a factory, but on the farm the foreman was the pace-setting horseman.
haflin	young man not yet entitled to a man's wage
hairst(-man)	harvest; man engaged for the harvest
hale	whole
hash	hurry, rush
hine	far, long
hyow(in)	hoe(ing)
in bye	casual visit
kittle	tricky, tickle
kittlin	kitten
loon	boy
lows	to end a shift
lousin time	time to end a yoken: 11.30 and 6.00 in the 1930s
lythe	shelter
mealy Jimmy *or* **mealy pudding**	sausage made with oatmeal, onions and dripping
neep	turnip
orrabeast	single horse or extra horse, not one of a pair
orraman	farm servant who is not a grieve, cattleman or tractorman. He may do any job as required.
peesie, peesieweep	lapwing or green plover
plank (down)	put money down on the counter with a flourish

ploo	plough
preen	pin
pu	pluck
puin neeps	plucking, topping and tailing turnips
quine	girl
ragnail(s)	bits of loose nail or skin on feet
rape(s)	home-made ropes of straw
reeshle	when corn is really dry it will rustle in a characteristic way
roup	displenish sale
ruck	stack of straw or hay
ruggit	destroyed by pulling – nasty youths used to destroy birds' nests by ruggin them
shaw	stem of potatoes or leaf of turnips
shelt	small horse like a Shetland pony
shim	implement for weeding the sides of drills
single; singling	to hoe; hoeing turnips so as to get rid of weeds and leave single turnip plants rather than clumps
sma'hudder	smallholder, farmer of a small farm or smallholding
smeddum	vigorous commonsense
sook	suck
speir (about)	ask, ask around about. Farm servants who the farmer wanted to stay would be 'speired' a few days before the term – invited to stay on.
stirkie	weaned calf of either sex between about four months and a year old
stook	eight or ten sheaves built together on end to help them dry
stot	steer
straik	one round of an operation – when carting neeps from field to byre, one straik was one trip to the field and back
Suntie	Santa Claus
suppy	small amount of
sweel; sweel the byre	swill; splash the byre with buckets of water to clean it
sweir	hard, difficult
tackety boots	universally worn farming boots with their soles covered with nobbly headed nails to give grip and make them hard-wearing

tag short leather strap for bashing discipline into children. It had two tails and when the strap hit the offender's hand the tails nipped the soft skin on the wrist.

takye the game of tig

tapner hooked knife for plucking, topping and tailing turnips

tattie potato

teem to empty

teuchter unsophisticated rural person

there anent involved

there's aye some water far the stirkie droons there's no smoke without fire

thirled breeding sheep are thirled to the hill they are born on and won't easily settle or do well anywhere else

thrash(ing) threshing. Separating the straw and chaff from the grain

tippling head attachment for a digger which allows loads to be tipped after raising them

ton the old Imperial ton in use before decimalisation in the 1970s. It consisted of 20 hundredweights of 112 lbs each (1 ton = 2,240 lbs).

tonne the decimal tonne (or short tonne) is 1,000 kilos (1 tonne = 2,200 lbs)

toonkeeper person left in charge and with all the essential work to do when all other staff are off-duty

tory worm grubs which can play havoc with young cereal plants

wid would

wreath snowdrift

yalla yite yellow hammer

yoken shift

yokin time the start of a shift when the tractor or horse has to be yoked to whatever machine is to be used

Index